Helga Krinzinger

The role of multi-digit numbers in the development of numeracy

Helga Krinzinger

The role of multi-digit numbers in the development of numeracy

Südwestdeutscher Verlag für Hochschulschriften

Imprint

Any brand names and product names mentioned in this book are subject to trademark, brand or patent protection and are trademarks or registered trademarks of their respective holders. The use of brand names, product names, common names, trade names, product descriptions etc. even without a particular marking in this work is in no way to be construed to mean that such names may be regarded as unrestricted in respect of trademark and brand protection legislation and could thus be used by anyone.

Publisher:
Südwestdeutscher Verlag für Hochschulschriften
is a trademark of
Dodo Books Indian Ocean Ltd., member of the OmniScriptum S.R.L Publishing group
str. A.Russo 15, of. 61, Chisinau-2068, Republic of Moldova Europe
Printed at: see last page
ISBN: 978-3-8381-2625-8

Zugl. / Approved by: Aachen, RWTH Aachen, Diss., 2010

Copyright © Helga Krinzinger
Copyright © 2011 Dodo Books Indian Ocean Ltd., member of the OmniScriptum S.R.L Publishing group

"If you take care of the syntax of a representational system, the semantics will take care of itself."

(Haugeland, 1985, p. 106)

The role of multi-digit number processing in the development of numerical cognition

Table of Contents

1 General Introduction _____ 9
   1.1   Can adult models be used to describe the development of numeracy? __ 10
   1.2   Is the transcoding process in children semantic or asemantic? _____ 11
   1.3   Are multi-digit number representations in children holistic or decomposed? _____ 12
   1.4   Do visual-spatial abilities explain individual (and gender) differences in multi-digit number processing in children? _____ 12
   1.5   General outline of the studies and specific research questions _____ 13

2 Study 1: Comparison of one-, two-, and three-componential models of early numeracy _____ 15
   2.1   Introduction _____ 15
      2.1.1   Adult models of numeracy _____ 15
      2.1.2   Models based on subtypes of developmental dyscalculia _____ 16
      2.1.3   Developmental stage models _____ 19
      2.1.4   Objectives of Study 1 _____ 20
   2.2   Methods _____ 21
      2.2.1   Participants _____ 21
      2.2.2   Dyscalculia test battery _____ 22
      2.2.3   Procedure and tasks _____ 23
      2.2.4   Modes of analysis _____ 25
      2.2.5   Structural models _____ 26
      2.2.6   Fit statistics _____ 28
   2.3   Results _____ 29
      2.3.1   Descriptive statistics and gender differences in raw scores _____ 29
      2.3.2   One-componential model _____ 31
      2.3.3   Two-componential model _____ 32
      2.3.4   Model fit comparisons between one- and two-componential models 34
      2.3.5   Three-componential model _____ 35
   2.4   Discussion _____ 37
      2.4.1   Evaluation of the three-componential model _____ 37
      2.4.2   Comparison of the one- and the two-componential model _____ 38

  2.4.3 Linking the two-componential model to previous theories of number processing _____ 40

 2.5 Conclusion _____ 41

3 Study 2: Differential linguistic effects on numerical tasks _____ 42

 3.1 Introduction _____ 42

  3.1.1 Language effects on the development of numeracy _____ 42

  3.1.2 Inversion effects on the development of numeracy _____ 44

  3.1.3 Objectives of Study 2 _____ 45

 3.2 Methods _____ 46

  3.2.1 Participants _____ 46

  3.2.2 Procedure and tasks _____ 47

  3.2.3 Analyses _____ 48

 3.3 Results _____ 48

  3.3.1 Writing Arabic numbers to dictation _____ 48

  3.3.2 Recognition of unit- and decade-digits _____ 50

  3.3.3 Subtraction _____ 51

 3.4 Discussion _____ 53

  3.4.1 Inversion effects between language groups _____ 53

  3.4.2 Specific vs. unspecific effects of inversion _____ 54

  3.4.3 Positive effects of inversion on Subtraction _____ 55

  3.4.4 No inversion effect on Recognition of unit- and decade-digits _____ 55

 3.5 Conclusion _____ 57

4 Study 3: Influence of gender equality on gender differences in numerical tasks _____ 58

 4.1 Introduction _____ 58

  4.1.1 Higher performance variability in males as a possible reason for gender differences in mathematics _____ 58

  4.1.2 Socio-cultural reasons for gender differences in mathematics _____ 59

  4.1.3 Fact retrieval advantage as a possible reason for gender differences in mathematics _____ 60

  4.1.4 Objectives of Study 3 _____ 61

 4.2 Method _____ 61

4.2.1	Participants and tasks	61
4.2.2	Gender Gap Index	61
4.2.3	Analysis	62
4.3	Results	62
4.3.1	Writing Arabic numbers to dictation	62
4.3.2	Recognition of unit- and decade-digits	63
4.3.3	Subtraction	64
4.3.4	Multiplication	65
4.4	Discussion	66
4.5	Conclusion	67

5 Study 4: What accounts for gender differences in multi-digit number processing? 68

5.1	Introduction	68
5.1.1	The spatial cognition hypothesis of gender differences in mathematical cognition	68
5.1.2	The psychobiosocial model of gender differences in mathematics	69
5.1.3	Objectives of Study 4	70
5.2	Methods	70
5.2.1	Participants	70
5.2.2	General procedure	71
5.2.3	Stimuli and task procedures	71
5.2.4	Model specification	74
5.2.5	Fit statistics	77
5.3	Results	77
5.3.1	Descriptive statistics	77
5.3.2	Longitudinal model	77
5.4	Discussion	82
5.4.1	Why are visual-spatial abilities important for the acquisition of multi-digit number processing?	82
5.4.2	Why is visual-spatial working memory capacity not a predictor for multi-digit number processing?	83
5.5	Conclusion	84

6 Study 5: Multi-digit number processing as predictor of multi-digit calculation performance and an exact number magnitude representation: a longitudinal structural equation model _____ 85

6.1 Introduction _____ 85
6.1.1 Models of number processing _____ 85
6.1.2 The role of multi-digit number processing in the development of an exact, linear number magnitude representation _____ 87
6.1.3 The role of multi-digit number processing in learning multi-digit calculation _____ 89
6.1.4 Developmental influences between multi-digit calculation performance and linearity of number magnitude representation _____ 90

6.2 Objectives of Study 5 _____ 90

6.3 Methods _____ 91
6.3.1 Participants and general procedure _____ 91
6.3.2 Stimuli and task procedures _____ 91
6.3.3 Model specification _____ 93
6.3.4 Fit statistics _____ 94

6.4 Results _____ 94
6.4.1 Descriptive Statistics _____ 94
6.4.2 The role of number symbol knowledge in the number range 0-100 for the linearity of number representation in the same range _____ 99
6.4.3 Longitudinal model _____ 99

6.5 Discussion _____ 103
6.5.1 Is flawless number symbol knowledge sufficient for an exact number representation? _____ 103
6.5.2 Is general base-10 knowledge as needed for multi-digit number processing a predictor for an exact number representation and for calculation skills in the range 0-100? _____ 104
6.5.3 Is an exact number representation predicting calculation skills or vice versa? _____ 106

6.6 Conclusions _____ 106

7	**General Discussion**	**108**
7.1	The core mechanism underlying children's multi-digit number processing: Base-10 understanding	109
7.2	Adult models are not adequate for the description of the development of numeracy	110
7.3	The transcoding process in children is at least mediated by semantic number representations	111
7.4	The decomposed aspect of multi-digit number representation is predominant during childhood	112
7.5	The role of visual-spatial abilities in multi-digit number processing	113
7.6	Implications for mathematics education and treatment of developmental dyscalculia	115
8	**References**	117
9	**Appendix**	129

The role of multi-digit number processing in the development of numerical cognition

1 General Introduction

The main aim of this thesis is to investigate the role of multi-digit number processing in the development of numeracy. Numeracy (i.e., the ability to reason with numbers) is in our modern society more important than literacy for employment rates and wages (Dowker, 2005). Therefore, a thorough understanding of the mechanisms which underlie the development of numerical cognition is of socio-cultural importance, as it could help informing math teachers and professionals training children diagnosed with developmental dyscalculia (DD) in order to elicit the optimal mathematical performance in our children.

Even though research on numerical cognition in children has increased during the last decades (for a review, see Ansari, 2008), the most difficult prerequisite for numeracy – namely the acquisition of the Arabic number system (Geary, 2000) – has been largely understudied. In the 1990ies, several studies have focused on language differences in the development of multi-digit number processing (e.g., Fuson & Kwon, 1991; Miura, Okamoto, Kim, Steere, & Fayol, 1993; Miller, Smith, Zhu, & Zhang, 1995), on the importance of place-value understanding for the execution of multi-digit calculation procedures (Baroody, 2003; Hiebert & Wearne, 1996; Resnick, 1992; Rittle-Johnson & Siegler, 1998), or on the central conceptual structures underlying multi-digit number processing (Case, 1996; Fuson, Wearne, Hiebert, Murray, Human, Olivier, Carpenter, & Fennema, 1997).

Yet, recent cognitive research in the domain of numerical cognition in children frequently ignores these older studies and is based on adult models of numeracy, usually the Triple-Code model by Dehaene (1992). The current focus in this domain lies on the distance effect as a marker for basic number representations as predictor for calculation ability (e.g., Ansari, 2008, for a review of latest findings and the most important research questions which should be faced in the future).

One exception is the research group around Liane Kaufmann and H.-C. Nuerk, who employ tasks tapping multi-digit number processing in children (e.g., Helmreich, Zuber, Pixner, Kaufmann, Nuerk, & Moeller, accepted; Kaufmann & Nuerk, 2005; Moeller, Pixner, Kaufmann, & Nuerk, 2009; Nuerk, Kaufmann, Zoppoth, & Willmes, 2004; Pixner, Moeller, Zuber, & Nuerk, 2009; Zuber, Pixner, Moeller, & Nuerk, 2009). These studies are all experimental in nature and focus on specific cognitive effects on specific numerical tasks.

Yet, as Cronbach has pointed out in 1975, experimental research is limited in its ability to detect relationships between different factors (which will likely play a role in the development of numerical cognition) which are not captured by the respective hypotheses. On the other hand, correlational research "accepts the natural range of variables, instead of shaping conditions to represent a hypothesis" (Cronbach, 1975, p. 124).

This thesis will try to shed some light on developmental mechanisms active in the typical acquisition of numeracy, which implies more general research questions than whether a specific reaction time effect found in adults is also present in children, and if so from which age onwards. Therefore, four out of five studies presented here will use correlational methods such as structural equation modelling.

1.1 Can adult models be used to describe the development of numeracy?

The most general question which should be answered here is whether adult models of numerical cognition also hold for describing its development. Several authors have pointed out that this is not necessarily the case (e.g., Ansari, 2010; Karmiloff-Smith, 1992, 1998; Kaufmann & Nuerk, 2005). Specifically, three assumptions have to be true in order to assume that adult models can account for the description of (disorders in) the development of numerical cognition (Ansari, 2010): First, performance profiles should not change over time. Second, similar neuronal structures should subserve numerical cognition in children and adults. Third, similarities in behavioural performance should imply equivalence in underlying neurocognitive mechanisms. Several studies, mostly on individuals with genetic disorders, have shown that these assumptions do not hold in every case (for an overview, see Ansari, 2010).

Yet, up to date nobody has ever tested empirically whether adult models can describe typical development of numeracy. In Study 1, we will do so.

1.2 Is the transcoding process in children semantic or asemantic?

A comparison of the two most influential adult models of numerical cognition (Triple-Code model, Dehaene, 1992; model by McCloskey, 1992, based on an older version by McCloskey, Caramazza, & Basili, 1985) in respect to their assumptions about multi-digit number processing leads to two other general questions: namely the nature of transcoding processes for multi-digit numbers and the nature of multi-digit number magnitude representation.

The Triple-Code model (Dehaene, 1992) distinguishes a module for the processing of verbally presented numbers, of numbers presented in the Arabic modality, and of the abstract, analogue magnitude representation of numbers. The latter is thought to resemble a spatially oriented number line with a continuum of small numbers on the left and large numbers on the right side of space (holistic magnitude representation). These modules can be impaired independently from each other, but are not described in the strict sense of cognitive modules (Fodor, 1983).

The transcoding of numbers from the verbal to the Arabic modality does not necessarily need semantic mediation of the magnitude representation (asemantic transcoding).

McCloskey (1992) differentiates between input- and output-processes for verbally presented and Arabic numbers as well, but he assumes a mandatory semantic mediation by the internal number magnitude representation (semantic transcoding). Furthermore, the abstract internal semantic representation is thought to rely on the base-10 structure of our number system and that the representations of large numbers are decomposed into its multiples of the power of ten (e.g., 317 would be represented as $3EXP\{2\} + 1EXP\{1\} + 7EXP\{0\}$; decomposed semantic magnitude representation).

The question of whether the transcoding process is semantic or asemantic in children has not yet attained a lot of attention. The only current developmental transcoding model (ADAPT; Barrouillet, Camos, Perruchet, & Seron, 2004) holds a purely asemantic and procedural view. Yet, the results of Studies 1, 2, 4, and 5 lend support to a different view.

1.3 Are multi-digit number representations in children holistic or decomposed?

As mentioned above, the two models by Dehaene (1992) and McCloskey (1992) differ not only in their predictions concerning the transcoding process of multi-digit numbers, but also in their conceptualizations of number magnitude representation. The Triple-Code model holds that number magnitude representations are holistic (comparable to an analogue mental number line), whereas the McCloskey-model advocates decomposed number representations mirroring the base-10 system.
Nuerk and Willmes (2005) review a broad range of different experimental studies in adults concerning this question and draw the conclusion of hybrid representations for large numbers including both holistic as well as decomposed aspects.
In children, this question has rarely been investigated. An experimental study by Nuerk and colleagues (2004) concluded that at least from second grade onwards, numbers are processed in a decomposed fashion and not holistically in a task where children had to compare two double-digit numbers. Yet, they concluded that children may use different strategies depending on task demand. Other studies by Nuerk and collaborators are also supportive for the decomposed view (Moeller et al., 2009; Pixner et al., 2009), but again it is not clear whether the decomposed processing of double-digit numbers observed in these studies reflects only specific strategies elicited by the respective tasks on the one hand or a general number representation underlying all kinds of numerical cognition in childhood on the other hand.
In this thesis, the very question will be again addressed with correlational research methods to explore whether the notion of decomposed number representations may explain performance patterns across a range of different numerical tasks in children (see Study 1 and Study 5).

1.4 Do visual-spatial abilities explain individual (and gender) differences in multi-digit number processing in children?

The last general research question addressed in this thesis is whether visual-spatial abilities are required for multi-digit number processing in children. This question is of considerable importance as we observed gender differences in a multitude of tasks using multi-digit number processing in the standardization sample of the dyscalculia test-battery TEDI-MATH (Kaufmann, Nuerk, Graf, Krinzinger, Delazer, & Willmes, 2009). Adult neuropsychological patients with visual-spatial deficiencies frequently present with problems in the alignment of multi-digit Arabic numbers for written calculations (spatial acalculia: Hartje, 1987; Strang & Rourke, 1985). Based on these observations in neuropsychological adult patients, Geary (1993) proposed a respective spatial subtype of DD (see also: Rourke, 1989; Rourke & Finlayson, 1978; Rourke & Strang, 1978). Yet, in a recent reflection on his original subtyping of DD, Geary (2010) stated that supportive evidence for this spatial DD subtype has been scarce. Apart from the descriptions of a spatial dyscalculia subtype with problems in processing multi-digit Arabic numbers, no study has yet investigated the role of general visual-spatial abilities for multi-digit number processing (but see Zuber et al., 2009, for the impact of visual-spatial working memory capacity on first graders' transcoding abilities). Only in other numerical domains such as analogue spatial representation of numbers (Bachot, Gevers, Fias, & Roeyers, 2005) or mathematical problem solving (Casey, Nuttal, & Benbow, 1995) the influence of visual-spatial abilities or problem solving strategies has been shown. Two studies even observed that visual-spatial abilities mediated gender differences in mathematical problem solving in adults (Casey, Nuttal, & Pezaris, 2001) and children (Rosselli, Ardila, Matute, & Inozemtseva, 2008). Therefore, a possible impact of visual-spatial abilities on the development of multi-digit number processing will be evaluated in Study 4.

1.5 General outline of the studies and specific research questions

In the following section, the five studies and their specific research questions will be briefly outlined.
Study 1 evaluates and compares a one-componential model based on developmental stage models of numeracy, a two-componential model based on the finding of gender-differences in all tasks tapping multi-digit number processing in the standardization sample of the German version of the TEDI-MATH (Kaufmann et al., 2009), and a three-componential model based on the Triple-Code model (Dehaene, 1992). As these models are tested in the same standardization sample, C-scores corrected for gender differences will be used as observed variables in the structural equation models. All models are

specified once for the whole sample of n = 411 children and once for the four different age groups separately, but simultaneously to test for developmental stability of the respective solutions.

In Study 2, the effect of a specific property of some (e.g., German and Flemish) verbal number systems, namely the inversion of units and decades in spoken and written number names, on the performance on different numerical tasks is tested in multiple groups of children from countries speaking either a language with inversion (Austria, Germany, and Flanders) or without inversion (Wallonia and France). Conducting item analyses, the specificity (affecting only double-digit numbers) or generality (affecting also other numbers) of possible inversion-effects will be analysed as well. This study is accepted for publication in the Journal of Cross-Cultural Psychology (Krinzinger, Grégoire, Desoete, Kaufmann, Nuerk, & Willmes, accepted).

Study 3 will explore the mediating effect of national gender equality on performance differences between primary school boys and girls (see Study 1) in different numerical tasks. This is important, as gender equality has been shown to be a predictor for the male advantage in PISA scores in mathematics (Guiso, Monte, Sapienza, & Zingales, 2008). Even though Study 3 is only based on data from four different countries (Austria, Germany, Belgium, and France), it is the first study to investigate this question in primary school children.

In Study 4, the relative impact of visual-spatial abilities, visual-spatial working memory capacity (see above), and subjective evaluation of and/or attitudes towards mathematics (Eccles & Jacobs, 1986) on individual (and gender) differences in the development of multi-digit number processing will be investigated. This will be done employing longitudinal structural equation modelling.

Finally, Study 5 evaluates the importance of multi-digit number processing for double-digit calculation skills and the exactness of number magnitude representations in the range 0-100. Furthermore, the hypothesis that number symbol knowledge is sufficient for an exact number representation (e.g., Verguts & Fias, 2004; Dehaene, 2007) is tested, and the reciprocal influence of calculation ability and exactness of number magnitude representation is analysed as well. Again, structural equation modelling will be applied to a longitudinal data set capturing development during the first years of primary school.

2 Study 1: Comparison of one-, two-, and three-componential models of early numeracy

2.1 Introduction

Cognitive processing models are helpful in describing the functioning of the human mind scientifically. In classical cognitive psychology, one prominent way to establish a model is to investigate the malfunctioning of specific aspects of performance after focal brain damage to inform cognitive models of normal functioning (e.g., McCloskey, Caramazza, & Basili, 1985). It is however questionable whether models based on findings in fully developed adult individuals can also help to describe the development of cognitive functions during childhood (Ansari, 2010; Karmiloff-Smith, 1992, 1998; Kaufmann & Nuerk, 2005), as adult models usually do not consider learning processes. In this study, we will first describe adult models in the area of numeracy (including both basic number processing and calculation). Second, we will review developmental models in the same field based on atypical development on the one hand and stage models of normal development on the other hand. Third, we will empirically test several of these models on an extensive data set using structural equation modelling.

2.1.1 Adult models of numeracy

In the 1980s and 1990s, scientific interest in the cognitive mechanisms and processes underlying numeracy increased. Two different comprehensive models of number processing and calculation findings were postulated and influenced most of the forthcoming research on numeracy.

One model trying to describe key aspects of numeracy was proposed by McCloskey (1992; based on an older model by McCloskey, Caramazza, & Basili, 1985, and extended to a multi-route model by Cipolotti & Butterworth, 1995). Based on neuropsychological case studies, this model intended to explain both normal and impaired numerical cognition and included input modules for encoding Arabic numerals and number words, a central module for an abstract, semantic representation of numerical magnitudes in the form of a base-10 system, modules for numerical operations / calculation, and output modules for the production of Arabic numerals and number words.

The so-called Triple-Code model was established by Dehaene (1992) and further extended to an anatomo-functional model shortly afterwards (Dehaene & Cohen, 1995). Based on findings from a large number of single-case studies of brain-damaged patients as well as brain imaging studies in healthy subjects, the latter model linked mental processes involved in number processing and mental arithmetic to neuroanatomical circuits thought to support these processes. The cognitive architecture proposed by the Triple-Code model contained three main mental representations of numbers, namely the visual Arabic number form, the verbal word frame, and the analogue magnitude representation. The Triple-Code model differs from McCloskey's model in several ways (see also 1.2 and 1.3). First, modules for input and output are not separated. Second, the magnitude representation is thought to be semantic and amodal as well, but based on an analogue representation like a mental number line (representing numerals spatially ordered from left to right). Third, in the Triple-Code model no separate module for calculation is proposed, but different arithmetic operations are thought to depend on different representational formats, namely multiplication on the verbal word frame, subtraction on the magnitude representation, and addition on both. Finally, the McCloskey model (1992) is unidirectional and completely serial, whereas in the Triple-Code model (1992) all three internal representations are thought to interact with each other.

2.1.2 Models based on subtypes of developmental dyscalculia

At about the same time, the first attempts to utilize neuropsychological methods like single case studies for research on the development of numeracy appeared. For example, Temple found a double dissociation in two dyscalculic children between impairments in arithmetic fact retrieval and in arithmetic procedures. Consequently, she concluded that like in acquired acalculia/dyscalculia, also in DD double dissociations emerge which may be considered to reflect a modular organization of mental processes (1991).

Shortly afterwards, Geary (1993) offered a subtyping of children's mathematical disabilities based on classical neuropsychological syndromes of number processing deficits. The first subtype he described is characterized by difficulties in the representation or retrieval of arithmetic facts from semantic memory (Badian, 1983; Ashcraft, Yamashita, & Aram,

1992), similar to the neuropsychological syndrome of anarithmetria (Hecaen, Angelergues, & Houillier, 1961; McCloskey, Aliminosa, & Sokol, 1991; Sokol, McCloskey, Cohen, & Aliminosa, 1991). Yet, anarithmetria is also often characterized by problems in executing arithmetic procedures as well. Based on the double dissociation described by Temple (1991), Geary (1993) defines the deficit in procedure execution as a subtype of its own. The third subtype he proposed is mainly manifested by problems in the visual-spatial representation of numerical information (Rourke, 1989; Rourke & Finlayson, 1978; Rourke & Strang, 1978). This subtype is thought to be similar to the syndrome of spatial acalculia, characterized by misalignment of numbers in multicolumn arithmetic problems, number omissions, number rotations, misreading arithmetical operation signs, and difficulties with place value and decimals (Hartje, 1987; Strang & Rourke, 1985). The syndromes of alexia and agraphia for numbers (Cohn, 1961, 1971; Temple, 1989) were described as well. Yet, according to Kosc (1974) these syndromes may occur in children, but not as often as spatial acalculia and anarithmetria. Some years later, Geary (2004) proposed the same three subtypes of DD, but related them to difficulties in specific aspects of working memory (Baddeley, 1986): a procedural subtype caused by a delay in the development of central executive functions, a semantic memory subtype due to impairments in the phonological loop, and a visual-spatial subtype with problems in visual-spatial working memory capacity (see also Kaufmann, 2002; Kaufmann et al., 2004). Recently, Geary (2010) related findings of the last years to his original proposals of dyscalculia subtypes and stated that the originally proposed procedural and retrieval deficits have been frequently supported, but their underlying processes and mechanisms are more complicated than originally suspected. Furthermore, an additional number-sense deficit has been identified and should be seen as a specific subtype (e.g., Landerl, Bevan, & Butterworth, 2004). Finally, he claimed that support for the proposed visual-spatial deficit has been mixed, but that this could either be the case because it is less frequent than the other deficits (but see Bachot et al., 2005), or because the tasks used to assess dyscalculia may not require these competencies.

During the last decade, at least two subtyping systems for DD explicitly based on the Triple-Code model (Dehaene, 1992) have been published.

The first was based on the examination of a large sample of children and differentiated three potential subtypes apart from a subclinical group, namely a verbal subtype (characterized by high comorbidity rates with ADHD and reading disorders), an Arabic subtype (supposedly associated with visual-spatial problems), and a pervasive subtype with problems in all mathematical tasks, probably due to difficulties in the magnitude representation of numbers (von Aster, 2000).

Another recent respective subtyping system for DD was put forward by Wilson and Dehaene (2007). They proposed that DD may be caused by a core deficit (similar to a phonological processing deficit in dyslexia), which either should be due to deficient magnitude representations per se (i.e., deficient numerosity) or to deficient mappings between magnitude representations and symbolic representations of numbers. They also proposed three other possible subtypes which are quite similar to the subtypes described by Geary (2010): a subtype with deficits in the verbal symbolic representation, a subtype characterized by executive dysfunctions, and a subtype with spatial attention deficits (manifested in deficient subitizing and quantity manipulation). Yet, it is hard to separate their descriptions of the first possible core deficit (deficient magnitude representation) and the spatial subtype as well as of the second possible core deficit (deficient mapping of magnitude and symbolic number representation) and the verbal subtype.

In summary, most of the models presented by now differentiate between a verbal component important for fact retrieval and processing of number words and a (spatial) Arabic component that is important for the processing of multi-digit numbers. Moreover, they all describe a third and/or fourth component either important for executing arithmetic procedures (operating with numbers) or a component for (spatial) magnitude representation important for comparing and manipulating numerical information.

An overview about multi-componential models is provided in Table 2.1. Importantly, all of the aforementioned models are largely hypothetical in nature since the formulation of potential subtypes has been based on descriptive post-hoc classifications of samples or single cases reported in the literature. Thus, to the present, no empirically validated developmental model of numerical cognition informed by deficit patterns exists.

Table 2.1: Overview about multi-componential models of numerosity

	Verbal component	Arabic component	Magnitude component	Procedure execution component
McCloskey, 1992	x	x	x	x
Dehaene, 1992	x	x	x	
Temple, 1991	x			x
Geary, 1993	x	x		x

Geary, 2004	x	x		x
Geary, 2010	x	x	x	x
von Aster, 2000	x	x	x	
Wilson & Dehane, 2007	x		x	x

2.1.3 Developmental stage models

A completely different approach to modelling the development of numeracy is to describe different cognitive stages children have to pass through until they finally reach mature number processing and calculation abilities. Piaget (1952) was the first to describe stages or levels of development in the construction of number and arithmetic operations, which he believed to be closely related to the development of logic. His notion of basic central conceptions which he thought to evolve during the development of numerical cognition has been further elaborated in two slightly different models.

The more general theory by Case (1996) emphasizes that central conceptual structures represent children's core knowledge in a specific domain, and that they can be applied to a broad range of tasks. These conceptual structures are thought to become increasingly complex during development (from unidimensional to bidimensional to integrated bidimensional thought) by a reciprocal process of general conceptual insights and more specific task understanding. In the number domain, these stages are thought to incorporate the following aspects: In the first stage, analogue magnitude understanding and counting knowledge are merged together into a mental number/counting line (unidimensional). In the second stage, two mental number lines are linked to work in tandem (e.g., for ones and tens; bidimensional). In the third stage, relations between different mental number lines are established and refined (integrated bidimensional; Okamoto & Case, 1996).

Similarly, Fuson and her colleagues differentiate between different conceptual structures of double-digit numbers that evolve stepwise during development. First children master a unitary counting-string based representation. Second, a non-integrated two-dimensional concept will be acquired. Third, they succeed in integrating the two dimensions of units and decades into one bi-dimensional concept of double-digit numbers (Fuson et al., 1997). This model explicitly predicts that children's conceptual structures for double-digit numbers will influence their abilities in other numerical tasks as well, most importantly in calculation (Fuson et al., 1997).

A more recent model by von Aster and Shalev (2007) puts the different number representation formats described by the Triple-Code model (Dehaene, 1992) into a developmental context and describes their acquisition during childhood: From birth onwards, a core system is assumed to incorporate an analogue, approximate magnitude representation and subitizing. Preschool children will learn the verbal number system. During the first years of formal schooling, the Arabic number system is acquired which is the foundation of written arithmetic. Within a final developmental stage the model postulates the formation of a linear mental number line for exact number representations. These developmental stages are seen as elaborations of the basic core system, or representational redescriptions of the former according to Karmiloff-Smith (1992, 1998).

All these developmental stage models have in common that they consider numeracy as one cognitive component which is enriched in content during childhood. The same unitary view on numeracy underlies (i) the employment of intelligence tests incorporating scales for mathematical ability like the subscale quantitative reasoning of the Stanford-Binet IQ Test (Becker, 2003), (ii) using standardized achievement tests (e.g., the Stanford Achievement Test [SAT] Series; the Wechsler Individual Achievement Test – Second Edition [WIAT-II] by Wechsler, 2001) for cognitive research on numerical achievement in children (e.g., Geary, Hamson, & Hoard, 2000; Meyer, Salimpor, Wu, Geary, & Menon, 2010), (iii) giving one single grade for the school subject mathematics, and even (iv) diagnosing DD based on the total score of a dyscalculia test battery.

Nonetheless, such simplified approaches and overly crude conceptualizations of arithmetic skills fail to fully acknowledge the multi-componential and complex nature of numerical cognition (Kaufmann & Nuerk, 2005; Rubinsten & Henik, 2009; Raghubar, Barnes, & Hecht, 2010) and consequently, bear the danger to yield misleading and even wrong conclusions.

2.1.4 Objectives of Study 1

The standardization of the dyscalculia test battery TEDI-MATH (Kaufmann et al., 2009) comprised of different subtests in large samples of primary school children from different age groups allowed us to test some of the above described models empirically and

compare the respective model fit statistics. In the following section, we will briefly describe our rationale for depicting specific models of numerical cognition:
1) According to developmental stage models (e.g., Piaget, 1952; von Aster & Shalev, 2007) and studies with an implicit unitary view on numerical cognition, we will define a single-component model.
2) Based on the most influential multi-componential model (the Triple-Code model, Dehaene, 1992) and similar models from subtyping of DD (e.g., von Aster, 2000), we will define a three-componential model with a verbal, an Arabic, and a magnitude component of number processing.
3) Furthermore, we will define a two-componential model which is based on an important finding derived from the standardization process of the test battery, namely that all subtests using multi-digit numbers showed considerable gender differences in favour of boys (see also Zuber et al., 2009, for gender differences in writing multi-digit numbers in first graders). Therefore, we will distinguish a multi-digit number processing component as opposed to a calculation component. It is important to note that the distinction between a calculation component and a multi-digit number processing component is in line with the two representational modules separate from input or output processes described in McCloskey's model, namely a calculation module and a module for a base-10 semantic magnitude representation.

All three models will both be tested for the whole standardization sample of the TEDI-MATH (Kaufmann et al., 2009) as well as for the four age groups separately, but simultaneously. The latter analyses will allow us to test whether the factor loadings are comparable in size for different age groups, which would indicate stable cognitive patterns at least within the developmental span under investigation.

2.2 Methods

2.2.1 Participants

In the present study, the data of four age groups from the standardization sample of the dyscalculia test TEDI-MATH (Kaufmann et al., 2009) were analyzed. The youngest children attended first grade second semester (1_2), the next group second grade first semester (2_1), the second oldest second grade second semester (2_2), and the oldest group third grade first semester (3_1). In each age group, at least 50 boys and 50 girls were tested. Children were recruited from regular public primary schools from three different regions/states in Germany (Berlin, Brandenburg, Nordrhein-Westfalen) and two regions/states in Austria (Tyrol, Vienna). Written and informed consent was obtained from school officials as well as from children's caretakers. Sample characteristics can be obtained from Table 2.2.

Table 2.2: Sample characteristics of the four age groups: first grade second semester (1_2), second grade first semester (2_1), second grade second semester (2_2), and third grade first semester (3_1)

Age group	sex	n (%)	mean age in months (sd)
1_2	boys	50 (50)	90 (5.1)
	girls	50 (50)	89 (4.8)
	overall	100	89 (4.9)
2_1	boys	52 (50)	94 (5.1)
	girls	52 (50)	94 (4.2)
	overall	104	94 (4.7)
2_2	boys	53 (51.5)	101 (5.8)
	girls	50 (48.5)	100 (4.5)
	overall	103	100 (5.2)
3_1	boys	50 (48.1)	107 (5.1)
	girls	54 (51.9)	105 (6.3)
	overall	104	106 (5.8)

2.2.2 Dyscalculia test battery

The multicomponential dyscalculia test-battery TEDI-MATH was originally developed in French (Van Nieuwenhoven, Grégoire, & Noël, 2001) and translated into Dutch (Grégoire, Noël, & Van Nieuwenhoven, 2004), Spanish (Grégoire, Noël, & Van Nieuwenhoven, 2005), and German (Kaufmann et al., 2009). It has been tested for conceptual accuracy and clinical relevance in previous studies (e.g., Stock, Desoete, & Roeyers, 2009). Good psychometric properties were demonstrated for the German version (Kaufmann et al., 2009). This renders the TEDI-MATH very useful for cross-cultural studies on the development of numerical cognition (for language effects, see Krinzinger, et al., accepted, and Study 2).

2.2.3 Procedure and tasks

The full test battery was always administered individually to each child by a trained tester (usually a psychology or medical student) in a separate and quiet room in their schools. A testing session was only started if it could be completed on the respective day. Children were rewarded with a small present for participation. All items were scored with 1 for a correct and 0 for an incorrect answer. For some subtests, more items were used compared to the French version (Van Nieuwenhoven et al., 2001). Each subtest was terminated after five consecutive errors. In this case, all other subsequent items of the respective subtest were scored as incorrect. In the present study, only subtests without extreme ceiling effects in all four age groups were used. Raw scores are the number of correctly solved items. For all subtests used in this study, C-scores (with a mean of 5 and a standard deviation of 2) were calculated for each age group separately. In the following, the nine subtests used in this study will be briefly presented.

Magnitude comparison of Arabic numbers (MCA)
Children were presented with 18 pairs of Arabic numbers (4 single-digit, 6 double-digit, 4 three-digit, 4 four-digit) on a sheet of paper each and asked to point to the larger number.

Magnitude comparison of number words (MCW)
Children were asked whether the first or the second number of 20 orally presented number word pairs (4 single-digit, 2 mixed single- and double-digit, 6 double-digit, 4 three-digit, and 4 four-digit) was larger.

Transcoding – writing Arabic numbers to dictation (TW)
Children were asked to write down 28 different Arabic numbers (3 single-digit, 9 double-digit, 8 three-digit, and 8 four-digit numbers) that were presented verbally one by one by the tester.

Transcoding – reading Arabic numbers (TR)
Children were asked to read aloud 28 different Arabic numbers (3 single-digit, 9 double-digit, 8 three-digit, and 8 four-digit numbers) that were presented one by one on a sheet of paper. The items for TW and TR were not identical, but as similar as possible concerning difficulty.

Additive decomposition (AD)
Children were presented with a sheet of paper showing two meadows with the Arabic numbers 2 and 4 written on them. Then they were told, "A farmer has two meadows and six sheep. He puts two sheep on this meadow and four sheep on this meadow". Next, they were presented with two empty meadows on another sheet and asked, "How can he divide his six sheep differently between his two meadows?" All stated decompositions were noted, but on this first trial a maximum of two points could be gained. After the children stopped giving new decompositions, the task was repeated with eight sheep, and four more points could be gained by giving at least as many valid decompositions of the number eight. Hence, on this subtest a maximum of six points could be reached.

Subtraction (S)
This subtest was administered by showing the children one of 15 single- and double-digit subtraction problems after the other on a sheet of paper presented in the Arabic modality, asking them to solve the items mentally as quickly and accurately as possible and to give the correct answers verbally. Six items require a borrow-procedure to be solved correctly.

Multiplication (M)
This subtest was administered by showing the children 14 simple multiplication problems presented as Arabic numbers one after the other on a sheet of paper. All operands were smaller than ten, and three items were rule-problems (e.g., 8 x 0). Similar to the subtraction task, the examiner asked the child to solve them mentally as quickly and accurately as possible and to give the correct answers verbally.

Word problems (WP)
Children were shown 12 written word problems (with results and operands up to 20), each one on a separate sheet of paper. All items required either addition or subtraction. The tester also read the problems aloud and repeated them once if asked to. Children ought to give the correct answers verbally.

Arithmetic concepts (AC)
Children were presented with eight pairs of complete (e.g., 29+66=95) and incomplete (e.g., 66+29=?) arithmetic computations (without result) using only double-digit operands and results presented in the Arabic modality on a separate sheet of paper each. Children were asked not to calculate but to state whether the first computation helps in solving the second one. Moreover, children were asked to justify their answers. The arithmetic principles included in this subtest were commutativity of addition operators, of addition and subtraction, and of addition and multiplication.

2.2.4 Modes of analysis

To examine the respective fits for the one-, two-, and three-componential models specified, we applied confirmatory factor analysis (using AMOS 7.0). Latent variables can be interpreted as the commonality shared between all observed variables loading on the respective factors. In our case, each component constituted a latent variable, and the age-group specific C-scores of the subtests were used as observed variables.

We did not use raw scores, because for all subtests employing multi-digit number processing (both magnitude comparison and both transcoding tasks) considerable gender differences were observed (see 2.3.1), which were levelled out by the C-scores (Kaufmann et al., 2009). Therefore, the gender differences which motivated our proposal of a two-componential model can not influence the correlation patterns between tasks.

We used the maximum-likelihood estimation method (Hoyle, 1995) based on the correlation matrices for the observed variables (see Tables A2.1 and A2.2 in the Appendix). All models were tested once for the whole sample and once for all four age groups simultaneously (with age group taken as group variable) to test for developmental stability of factor loadings.

2.2.5 Structural models

One- and two-componential model
In the one-componential model, all subtests were assumed to load on a single latent variable which we called mathematical ability (Math) (see Figure 2.1). In the two-componential model, the subtests loaded on either one of the following two latent variables: multi-digit number processing (MuDi) or calculation (Calc) (see Figure 2.2).
In order to specify a nested one-componential model, the correlation between the two components was set to one in the first case. This yielded practically one latent variable in this model (Math) and allowed for a direct statistical comparison between the one-componential model (correlation parameter between components fixed at 1) and the two-componential model (correlation parameter between components allowed to vary freely; see Figure 2.2).

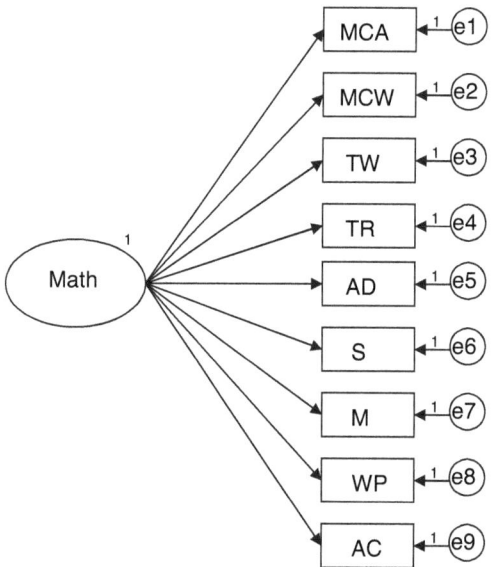

Figure 2.1: Model specification of the one-componential model

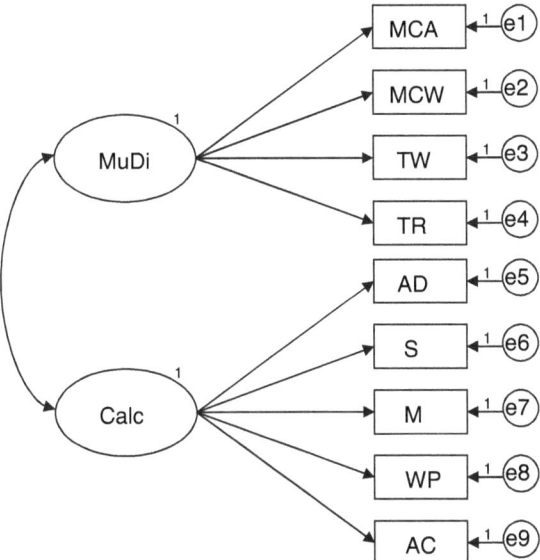

Figure 2.2: Model specification of the two-componential model

Three-componential model

In the three-componential model, the three latent variables mirroring the representational codes in the Triple-Code model, verbal processing, Arabic processing, and magnitude processing, were assumed to affect all subtests for which processing numbers in the respective modality is expected for either input, output, or internal processing (see Figure 2.3).

The component Arabic processing (Arab) was assumed to load on the subtests MCA, TW, TR, S, M, and AC, the component verbal processing (Verb) to load on the subtests MCW, TW, TR, AD, S, M, and WP. The component magnitude processing (Magn) was assumed to be relevant for processing the items in subtests MCA, MCW, AD, S, WP, and AC. Correlations between the three components were allowed.

As executing calculation procedures can be seen as a subset of abilities supported by the component Magn which is also responsible for manipulating quantities, a respective procedural component could be conceptualized as a component nested in Magn. If this notion is correct, the regression weights of MCA and MCW (which do not require the execution of calculation procedures) on Magn should be redundant and setting them to zero should increase model fit.

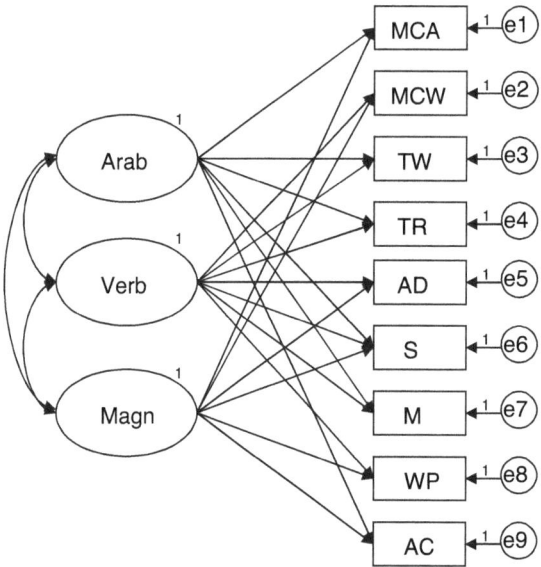

Figure 2.3: **Model specification of the three-componential model**

2.2.6 Fit statistics

In order to evaluate the goodness-of-fit of the sample data to our models, we used the chi-square test statistic of model fit (X^2, Jöreskog, 1969) as well as the Adjusted Goodness-of-Fit Index (AGFI: Jöreskog & Sörbom, 1984; Shevlin & Miles, 1998), the Comparative Fit Index (CFI: Bentler, 1992), the Root Mean Square Error of Approximation (RMSEA: Browne & Cudeck, 1993), and the Akaike Information Criterion (AIC: Akaike, 1987). For a good model, the ratio of X^2 divided by its associated degrees of freedom has to be less than 2 conventionally, CFI and AGFI larger than 0.95, RMSEA smaller than 0.05, and the AIC for the model chosen should be the smallest value among the alternative models considered.

2.3 Results

2.3.1 Descriptive statistics and gender differences in raw scores

As can be seen in Table 2.3, considerable gender differences were found for the raw scores of our samples in favour of boys.

Table 2.3: Means and standard deviations (in parentheses) of raw scores for all subtests separately for girls and boys as well as t-statistics for each age group (with Cohen's effect size d for significant gender differences) for each age group

Subtest		1_2	2_1	2_2	3_1
MCA	girls	12.4 (3.2)	12.6 (2.9)	15.6 (2.8)	17.1 (1.4)
	boys	14.2 (3.7)	15.6 (3.1)	16.4 (2.4)	17.2 (1.3)
	t	-2.6**	-5.1***	-1.7(*)	-0.4
	Cohen's d	0.54	1.04		
MCW	girls	14.9 (3.0)	15.3 (2.8)	18.0 (2.6)	19.3 (2.3)
	boys	16.8 (3.5)	18.0 (3.0)	19.1 (2.5)	19.7 (2.2)
	t	-2.9**	-4.7***	-2.2*	-0.9
	Cohen's d	0.61	0.94	0.43	
TW	girls	14.8 (5.1)	16.8 (4.1)	22.3 (4.3)	26.0 (2.9)
	boys	19.3 (7.0)	22.5 (5.4)	24.0 (5.0)	26.9 (2.0)
	t	-3.7***	-6.1***	-1.8(*)	-1.7(*)
	Cohen's d	0.74	1.25		
TR	girls	17.5 (5.3)	18.9 (3.9)	23.6 (4.1)	26.6 (2.2)
	boys	21.5 (5.6)	24.3 (4.6)	25.3 (4.2)	27.4 (1.4)
	t	-3.6***	-6.5***	-2.1*	-2.0*
	Cohen's d	0.73	1.28	0.42	0.40
AD	girls	5.0 (1.4)	5.0 (1.4)	5.5 (1.3)	5.6 (1.0)
	boys	5.4 (1.4)	5.2 (1.4)	5.8 (0.5)	5.7 (0.8)
	t	-1.3	-0.8	-1.8(*)	-0.3
	Cohen's d				
S	girls	8.1 (2.1)	9.8 (2.1)	12.9 (1.9)	13.0 (2.4)
	boys	8.8 (3.1)	10.8 (2.6)	12.7 (2.2)	13.3 (1.7)
	t	-1.4	-2.3*	0.7	-0.8
	Cohen's d		0.45		
M	girls	5.5 (4.4)	7.6 (4.3)	13.0 (1.3)	12.7 (1.7)
	boys	6.2 (4.3)	9.4 (3.6)	13.0 (1.9)	12.5 (1.7)
	t	-0.8	-2.3*	0.1	0.7
	Cohen's d		0.42		
WP	girls	8.6 (2.3)	8.6 (2.1)	9.7 (1.7)	10.1 (1.9)
	boys	8.4 (2.4)	9.5 (2.2)	9.7 (2.1)	10.2 (1.6)
	t	0.3	-2.2*	-0.2	-0.3
	Cohen's d		0.44		
AC	girls	3.7 (2.0)	4.5 (1.9)	5.4 (1.5)	6.5 (1.7)
	boys	3.6 (1.7)	5.0 (1.7)	5.3 (1.9)	6.5 (1.5)
	t	0.4	-1.3	0.3	0.1
	Cohen's d				

(*) $p < .10$; * $p < .05$; ** $p < .01$; *** $p < .001$

In order to prevent systematic diagnostic biases, we corrected the raw data for gender differences in each case a performance difference between boys and girls for a subtest in a specific age group was significant and at least two raw score points on average. We adjusted the gender correction according to the actual raw score distributions of boys and girls: Whenever the rounded raw score difference corresponding to a specific percentile was two, four, or six, half of the difference was subtracted from the boys' and added to the girls' scores. These corrected raw scores were then used for the calculation of C-score norm data for the joint sample of boys and girls per age group. This was the case for the subtests MCA, MCW, TW, and TR for children attending first grade second semester (1_2) and second grade first semester (2_1) as well as TW and TR for second grade second semester (2_2).

In the present study, we used the C-scores and not the raw scores as observed variables to avoid a systematic gender bias which could otherwise artificially strengthen the relation between some subtests and weaken the relation between others. Table 2.4 shows the descriptive statistics for the C-scores per subtest and each age group. The correlations between C-scores per age group can be obtained from Tables A2.1 and A2.2 in the Appendix.

Table 2.4: Descriptive statistics of C-scores for all subtests for each age group

Subtest		1_2	2_1	2_2	3_1
MCA	Mean (Sd)	4.9 (1.8)	5.1 (2.0)	4.8 (1.6)	4.8 (1.6)
	Range	0-8 [a]	2-8 [a]	1-7 [a]	0-6 [a]
MCW	Mean (Sd)	4.9 (1.8)	5.0 (2.0)	4.9 (1.9)	4.7 (1.6)
	Range	0-8	0-8 [a]	0-8 [a]	0-6 [a]
TW	Mean (Sd)	4.9 (1.8)	4.9 (2.0)	4.8 (1.9)	4.8 (1.7)
	Range	0-8 [a]	1-8 [a]	0-8 [a]	0-6 [a]
TR	Mean (Sd)	5.0 (1.8)	5.2 (2.0)	4.8 (1.6)	4.9 (1.6)
	Range	0-8 [a]	1-8 [a]	1-7 [a]	0-6 [a]
AD	Mean (Sd)	4.7 (1.4)	5.0 (1.6)	4.9 (1.6)	4.5 (1.3)
	Range	0-6 [a]	0-6 [a]	1-6 [a]	0-5 [a]
S	Mean (Sd)	5.0 (1.8)	5.2 (1.9)	5.1 (1.9)	4.9 (1.8)
	Range	0-9	0-9	0-9 [a]	0-7
M	Mean (Sd)	4.9 (1.9)	5.0 (2.1)	5.0 (2.1)	5.0 (1.9)
	Range	0-10	2-10	1-9 [a]	0-7 [a]
WP	Mean (Sd)	5.0 (1.8)	4.9 (1.8)	5.2 (1.9)	5.1 (1.7)
	Range	0-8	0-8 [a]	0-8	0-7 [a]
AC	Mean (Sd)	4.9 (1.9)	4.9 (1.9)	4.8 (2.0)	4.9 (1.8)
	Range	0-9	1-9	0-9 [a]	0-7

[a] This parameter deviates substantially from a standard normal distribution (skewness or kurtosis divided by its standard deviation larger than 2).

2.3.2 One-componential model

Standardized factor loadings of the one-componential model for the whole sample of 411 children can be seen in Figure 2.4. All factor loadings were significantly different from zero ($p < .01$). The lowest regression weight was found for AC (.52), and the highest for TW (.86). The variance explained by the latent variable can be found above each respective observed variable. Modification indices recommended adding 18 error covariances between observed variables, suggesting that more systematic variance was present in the data than specified in the factorial structure of the one-componential model. Nonetheless, we generally refrained from adding error covariances to make fit indices better comparable between models because we lacked theoretically driven motivation to guide the modifications in this part of the model.

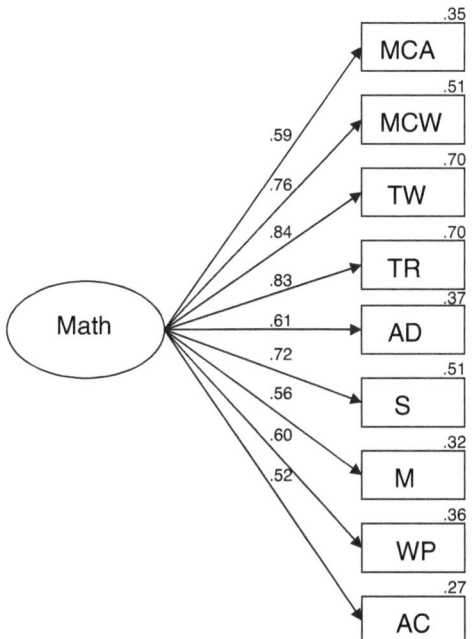

Figure 2.4: Standardized solution of the one-componential model (whole group)

The standardized regression weights of all subtests separately for each age group can be obtained from Table 2.5. Again, all factor loadings were significant ($p < .01$), the lowest loading was found for S (.40 for grade 3_1), and the highest for TW (.92 for grade 1_2). Modification indices recommended adding eight error covariances for grade 1_2, five for grade 2_1, five for grade 2_2, and seven for grade 3_1, again indicating that not all systematic variance in the data was captured by the model.

To test for equality of regression weights between age groups, we examined whether model fit would significantly decrease after equalizing the loading parameters for subtests for all four age groups. This constraint increased model fit significantly (difference in degrees of freedom: 21; difference in X^2: 40.5; $p = .007$).

Table 2.5: Standardized regression weights of the one-componential model for each observed variable (MCA: Magnitude comparison of Arabic numbers; MCW: Magnitude comparison of number words; TW: Transcoding: writing numbers; TR: Transcoding: reading numbers; AD: Additive decomposition; S: Subtraction; M: Multiplication; WP: Word problems; AC: Arithmetic concepts) and age group: first grade second semester (1_2), second grade first semester (2_1), second grade second semester (2_2), and third grade first semester (3_1)

Latent variable	Observed variables	1_2	2_1	2_2	3_1
Math	MCA	.57	.65	.58	.57
	MCW	.75	.77	.79	.74
	TW	.92	.75	.89	.79
	TR	.90	.83	.85	.78
	AD	.56	.57	.68	.66
	S	.70	.71	.64	.84
	M	.60	.70	.54	.40
	WP	.56	.65	.62	.66
	AC	.43	.47	.51	.74

2.3.3 Two-componential model

For the two-componential model, the solution for the whole sample of 411 children can be seen in Figure 2.5. Like for the one-componential model, all factor loadings were significant ($p < .01$), lowest for AC (.57) and highest for TW (.86). The covariance between the components MuDi and Calc was found to be .79. Only one error covariance was suggested by the modification indices.

The solutions for each age group separately can be obtained from Table 2.6. Covariances between the two components ranged between .76 and .82, and standardized regression weights from .46 (AC for group 1_2) to .93 (TW for group 1_2). Again, all values were significant ($p < .01$). Modification indices suggested only adding one error covariance for grade 1_2, one for grade 2_1, and two for grade 3_1.

Descriptively, factor loadings were higher for the subtests loading on Calc compared to their respective loadings for the one-componential solution.

Equalizing all regression weights and covariances between the components for all four age groups did not change model fit significantly (difference in degrees of freedom: 24; difference in X^2: 29.1; $p = .135$).

Table 2.6: Covariates between the latent variables multi-digit number processing (MD) and calculation (Calc) and standardized regression weights of the two-componential model for each observed variable (MCA: Magnitude comparison of Arabic numbers; MCW: Magnitude comparison of number words; TW: Transcoding: writing numbers; TR: Transcoding: reading numbers; AD: Additive decomposition; S: Subtraction; M: Multiplication; WP: Word problems; AC: Arithmetic concepts) and age group: first grade second semester (1_2), second grade first semester (2_1), second grade second semester (2_2), and third grade first semester (3_1)

Latent variable	Observed variable	1_2	2_1	2_2	3_1
MuDi ↔ Calc		.76	.82	.79	.80
MuDi	MCA	.57	.66	.60	.58
	MCW	.72	.78	.78	.77
	TW	.93	.75	.90	.85
	TR	.91	.87	.84	.81
Calc	AD	.60	.58	.72	.67
	S	.78	.76	.74	.84
	M	.70	.75	.61	.47
	WP	.71	.73	.67	.74
	AC	.46	.53	.57	.80

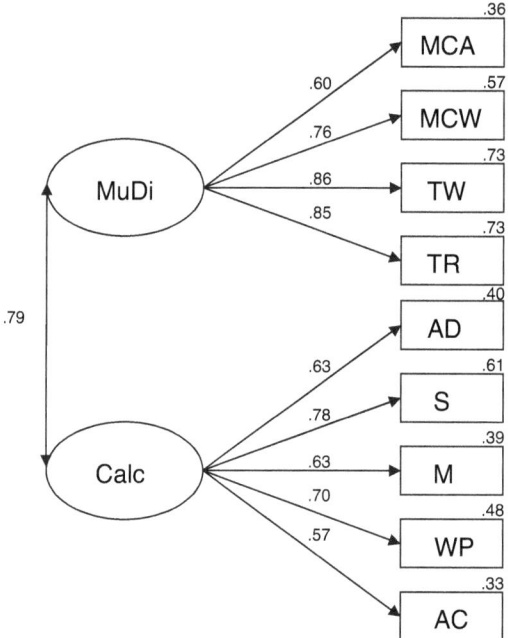

Figure 2.5: Standardized solution of the two-componential model (whole group)

2.3.4 Model fit comparisons between one- and two-componential models

The fit statistics of the converging models (one- and two-componential models respectively for the whole sample, four age groups simultaneously unconstrained, and four age groups simultaneously constrained in the way that regression weights are equalized between all groups) can be obtained from Table 2.7.

All three two-componential solutions reached fit indices conventionally considered as satisfactory for the ratio between X^2 and degrees of freedom, CFI, and RMSEA. The two-componential model for the whole sample reached a satisfactory value for the AGFI as well. The one-componential whole-sample solution and unconstrained as well as constrained solution for the four groups missed all cut-offs for a satisfactory fit index. All fit indices (most importantly, the AIC which can only be used for model comparison) were better for the two-componential solutions compared to the respective one-componential models.

The direct comparison between the one- and the two-componential models revealed that although the correlations between the two components were rather high (ranging from .76 to .82), model fit decreased significantly if they were set to 1 (difference in degrees of freedom: 1; difference in X^2: 110.8, $p < .001$ for the whole sample; difference in degrees of freedom: 4; difference in X^2: 115.7, $p < .001$ for the four age groups).

Table 2.7: Fits-statistics of the one- and the two-componential models for the whole group and all four age groups simultaneously unconstrained (all paths free) as well as constrained (regression weights equalized for all four age groups; X^2: chi-square; df: degrees of freedom)

	one-componential model			two-componential model		
	one group	four groups unconstrained	four groups constrained	one group	four groups unconstrained	four groups constrained
X^2	167	285	325	56	169	198
df	29	116	137	28	112	136
X^2 / df	5.8	2.5	2.4	2.0[a]	1.5[a]	1.5[a]
AGFI	.848	.777	.787	.951[a]	.869	.876
CFI	.903	.885	.872	.980[a]	.961[a]	.958[a]
RMSEA	.108	.060	.058	.050[a]	.035[a]	.034[a]
AIC	198	413	411	90[a]	305[a]	286[a]

[a] This value is conventionally considered to indicate satisfactory model fit.

2.3.5 Three-componential model

For the three-componential model, the covariance matrix for the whole sample of 411 children was not positively definite, implying that the solution was not admissible.
For the model simultaneously tested for four groups, the limit of 2000 iterations (predefined by AMOS 7.0) was reached. The model did not even converge with 5000 iterations. Therefore, also this solution was not admissible.
In order to diagnose the specification problem in our model, we examined the fit obtained when modelling each latent component separately. This analysis yielded three different sub-models. The model with the components Arab and Magn converged, but with poor fit indices (ratio of X^2 to degrees of freedom: 2.4; AGFI: .78; CFI: .93; RMSEA: .12), a correlation of .92 between the two components, and an implausible solution (e.g., non-significant regression weights from both components on MCA, S, and AC). The other two

sub-models (Verb and Arab; Verb and Magn) were unidentified and were therefore not admissible.

We also ran an exploratory principal components analysis with varimax rotation on the data of the whole sample and enforced a three principal components solution. The results for the whole sample after varimax rotation can be found in Table 2.8.

Table 2.8: Varimax rotated loadings of the subtests (MCA: Magnitude comparison of Arabic numbers; MCW: Magnitude comparison of number words; TW: Transcoding: writing numbers; TR: Transcoding: reading numbers; AD: Additive decomposition; S: Subtraction; M: Multiplication; WP: Word problems; AC: Arithmetic concepts) on three principal components (enforced three principal components solution)

Subtest	PC 1	PC 2	PC 2
MCA	.764	.187	.024
MCW	.780	.290	.050
TW	.810	.267	.163
TR	.818	.216	.227
AD	.139	.166	.941
S	.465	.525	.336
M	.262	.719	.049
WP	.229	.685	.322
AC	.181	.750	.039

The first factor (eigenvalue 4.4) accounted for 32% of the variance in the data, the second factor (eigenvalue 1.0) for 23%, and the third factor (eigenvalue 0.8) for 13% (overall: 68%). These results can be interpreted as indicating a principal component of multi-digit number processing (MCA, MCW, TW, and TR), a calculation component (S, M, WP, and AC), and a component comprising AD only.

In summary, the three-componential model failed to fit the data properly. Confirmatory factor analysis did not converge either when trying to fit the whole model or the submodels containing only two factors. Moreover, the principal components analysis revealed that the contribution of a third factor to the model is questionable and that is was not possible to distinguish a verbal and an Arabic component. The third component only had one variable (AD) with substantial loading and had only an eigenvalue clearly less than 1.

2.4 Discussion

The current study was designed to examine the validity of three possible cognitive models of early numeracy on a large data set of different numerical skills assessed in four age groups of primary school children. The main findings were that a one-componential model fitted less well compared to a model with the two latent variables/components Multi-digit number processing (MuDi) and Calculation (Calc) and that a three-componential solution derived from adult models and dyscalculia subtyping models did not converge. In the following, we will discuss our results in more detail.

2.4.1 Evaluation of the three-componential model

Research on the development of numerical cognition in primary school children has traditionally used an (at least implicit) unitary view or was based on adult models, most prominently the Triple-Code Model by Dehaene and Cohen (1995). The Triple-Code model proposes three different, but bidirectionally connected modules representing different number representations, namely an Arabic code for written numerals, a verbal code for number words and arithmetic fact retrieval like in multiplication, and an analogue magnitude code for comparing quantities and operating on them like in subtraction. Interestingly, at least two models of DD are explicitly linked to the Triple-Code model (von Aster, 2000; Wilson & Dehaene, 2007). To this end, we specified a three-componential model in which a verbal, an Arabic, and/or a quantitative component loaded on all subtests supposed to rely on processing of the respective numerical representation.

Surprisingly, the three-componential solution did neither converge for our whole sample nor for any of the four age groups separately. This means that the three components as proposed by the Triple-Code model may not adequately capture the variance in performance profiles of typically developing children when presented with a multitude of different numerical tasks. Rather, an exploratory principle components analysis with varimax rotation revealed similar results to the two-componential model. This finding does not lend support to the possibility that a procedural component as described by some dyscalculia subtyping approaches (Geary, 1993, 2004, 2010; Wilson & Dehaene, 2007) would be more adequate than (or in addition to) a magnitude component. Why should this be so?

A possible explanation is that the cognitive profiles typically found in development are not well suited to explain atypical cognitive processes and mechanisms characterizing DD and/or adult cognition.

Yet, when critically examining the subtyping models described in the introduction, it is rather surprising that the verbal subtype of dyscalculia is always only characterized by problems in arithmetic fact retrieval but not by coexisting problems in processing of number words. A plausible argument may be that these two skills need not be associated during development, even though both rely on retrieval of verbally presented numerical information. Rather, it is more plausible to speculate that during development verbal number processing may be tightly related to and not separable from Arabic number processing as two independent components, as the first is commonly thought to be a prerequisite of the latter (von Aster & Shalev, 2007). Furthermore, both verbal and Arabic multi-digit number processing share a base-10 representation (McCloskey, 1992), whereas for multiplication facts this representation may not be that relevant.

Hence, we suppose that early numeracy is a case in which adult models like the Triple-Code model (Dehaene & Cohen, 1995) can not be directly used for the description of typical development (Ansari, 2010; Karmiloff-Smith, 1992, 1998; Kaufmann & Nuerk, 2005).

2.4.2 Comparison of the one- and the two-componential model

As stated above, a unitary view on early numerical cognition can be found in all developmental stage models (Piaget, 1952; Case, 1996; Fuson et al., 1997; von Aster & Shalev, 2007) and in respective developmental studies using standardized achievement scores like the SAT (e.g., Geary, Hamson, & Hoard, 2000) or math grades solely. Developmental models of numerical cognition with a single component assume that performance related to different numerical skills can dissociate in children, but will be highly intercorrelated in the population.

Yet, the standardization data of the dyscalculia test TEDI-MATH disclosed two empirical observations that can not be explained by a unitary view (and neither by a three-componential model). First, we found highly significant gender differences in favour of boys in all subtests using multi-digit numbers irrespective of the task or the modality (see Table 2.3), which may be due to one underlying cognitive process needed for all tasks[1]. Second, even if correcting for these gender differences, all subtests using multi-digit numbers could be separated into one component which was not overlapping with a component comprising all calculation tasks as determined by a nonmetric multidimensional similarity scaling analysis (FSSA; Shye, 1985; Shye & Elizur, 1994; see Kaufmann et al., 2009). This nonmetric multi-dimensional similarity scaling method represents pairwise similarities of objects in a low-dimensional space in a way that objects (items) with high similarity are close to each other in space. If items are assessing similar aspects, the respective items should cluster together in space, and items containing different aspects should be spatially separable. The similarity measure used is the so called monotonicity coefficient – a measure of monotone relationship between items. Therefore, we contrasted a one-componential and a two-componential model with the factors Multi-digit number processing (MuDi) and Calculation (Calc) with data controlled for gender differences.

Our main result was that the fit of the two-componential model was satisfactory, no matter which fit index was used, whereas the fit of the one-componential model was poor. Furthermore, the one-componential solutions implied many more error covariances between observed variables as suggested by modification indices, indicating that much less systematic variance in the data was captured by the one-componential model specification. What is more, assuming equal loadings for all four age groups did not change model fit for the two-componential, but for the one-componential solution. This means that between the end of first and the middle of third grade, the observed correlational patterns of performance profiles were stable assuming two but not just one single cognitive component. Most importantly, the direct comparison revealed a significantly better fit for the two-componential model, even though the correlation between the two components was rather high.

In summary, a confirmatory factor analytic model assuming a cognitive component of multi-digit number processing and a calculation component is clearly superior compared to a unitary view on the development of numeracy.

---

[1] Analysing boys and girls as separate groups and equalizing regression weights between both groups did not change model fit neither for the two-componential solution (difference in degrees of freedom: 8; difference in X^2: 8.4, $p = .391$), nor for the one-componential solution (difference in degrees of freedom: 7; difference in X^2: 7.9, $p = .341$), indicating that no systematic differences in performance patterns between boys and girls existed.

2.4.3 Linking the two-componential model to previous theories of number processing

To summarize, our findings provide strong empirical evidence that numerical development is best conceptualized as comprising two cognitive components, namely multi-digit number processing and calculation. These two components found in our data closely correspond to two modules (for a base-10 semantic number representation and for calculation mechanisms) in the McCloskey model (1992).

It is important to note that these two modules of the McCloskey model are neither related to input nor to output processes, therefore being purely representational in nature.

Furthermore, our two-componential solution is also in line with the two types of conceptual knowledge Resnick (1982) differentiated in her longitudinal study, namely a concept of understanding the place-value principle of multi-digit numbers on the one side and concepts for calculation on the other side. Interestingly, two models of central conceptual structures in early numeracy also propose a mature concept of large numbers evolving from their inherent base-10 structure (Case, 1996; Fuson et al., 1997) and not from an analogue magnitude representation (Dehaene & Cohen, 1995). In fact, compared to an analogue magnitude representation, a base-10 concept underlying our multi-digit number processing component is much more likely, as thorough understanding of the base-10 system will definitely help in both transcoding and magnitude comparison of large numbers, but an analogue magnitude representation would only support the latter.

Based on our findings and the cited literature describing two distinct cognitive components of numerical cognition, which are all conceptual and/or representational in nature, we speculate that the acquisition of numeracy depends in general more on understanding central concepts than on rote memorization of calculation procedures (see also Baroody, 2003; Rittle-Johnson & Siegler, 1998) or carrying out purely syntactic transcoding algorithms (Barrouillet, Camos, Perruchet, & Seron, 2004), at least for typical development.

Importantly, we could also find a clear double dissociation in two children from the same age group between the multi-digit number component and the calculation component (Kaufmann et al., 2009), indicating that our two-componential model may also be useful for clinical practice. In particular, the two-componential solution may be relevant for both dyscalculia diagnosis (i.e., subtyping of DD) and rehabilitation planning, because children experiencing difficulties in the number processing domain – but not so in calculation –

require other types of intervention than children who display the opposite performance pattern.

2.5 Conclusion

Specifying a one-, a two-, and a three-componential model for a large data set on numerical skills in primary school children, we found that the three-componential solution based on the Triple-Code model (Dehaene & Cohen, 1995) did not converge. Furthermore, the two-componential solution, differentiating a multi-digit number processing and a calculation component, fitted better than a solution with only one component. Based on these findings we propose that models for adult numerical cognition can not be directly used to describe cognitive development. Rather, our findings strongly suggest that the two cognitive components of multi-digit number processing and calculation are at the core for the successful acquisition of early numeracy.

3 Study 2: Differential linguistic effects on numerical tasks

This study is accepted for publication in the Journal of Cross-Cultural Psychology (Krinzinger, Grégoire, Desoete, Kaufmann, Nuerk, & Willmes, in press).

3.1 Introduction

In our modern society, numeracy is becoming even more important than literacy for employment rates and wages (Dowker, 2005). In all modern societies, having sufficient numeracy means at least mastery of the Arabic number system and basic arithmetic operations. These numerical skills are taught in primary schools of all countries. Yet, this does not imply that the development of numeracy, which is a comparably young cultural invention (see: Dehaene, 1997), is the same in all cultures.

Furthermore, if performance differences in specific numerical skills are found between children of two different countries, these differences are often not unambiguously attributable to one cross-cultural distinction like differences in the number word systems. Rather, other cross-cultural differences (e.g. in math curricula) may be an important reason for performance differences as well. This means that if one wants to study the impact of a cultural difference like language effects on the development of numerical skills, possible confounding factors like curricular effects either have to be ruled out, or more than two different cultural groups have to be tested.

The aim of this study was therefore to investigate cross-cultural differences in the development of different numerical skills and to analyse whether eventual differences are due to linguistic or curricular effects.

3.1.1 Language effects on the development of numeracy

The most prominent culturally mediated difference in the development of mathematical skills is that highest performance can constantly be found in children from countries speaking a language in which the number word system is derived from Chinese (e.g., Singapore, Korea, Japan, Hong-Kong; TIMMS III, 1996; Singapore, Hong-Kong, China,

Japan; TIMMS 2007). In a comprehensive paper targeting exactly this cross-cultural difference between Asian and non-Asian children, Geary, Bow Thomas, Liu, and Siegler (1996) first reviewed relevant cross-national studies. All these studies argue that the reason for the Asian developmental advantage lies in the regularity and transparency of the Chinese-derived number word systems (e.g., Fuson & Kwon, 1991; Miura et al., 1993; Miller et al., 1995; see also Muldoon, Simms, Towse, Burns, & Yue, accepted). These number word systems are purely multiplicative and follow the base-10 Arabic notational system, which means that children only have to remember the digit names from 1 to 9 and the words for multipliers like 10, 100, and 1000 to be able to say every number word from 1 to 9999. On the other hand, children from Western cultures also have to remember the (sometimes irregular) names for teens and decade numbers (see: Comrie, 2005). Geary and co-workers (1996) then tested three possible reasons for the higher arithmetic competency of Chinese children compared to US-American children. They found that better base-10 understanding (due to the more transparent Chinese number word system), longer digit-span (due to shorter number words in Chinese), and higher number of math instructions in China – that is, curricular effects – were all significant mediators of the cross-cultural differences in arithmetic.

Several studies have supported the assumption that differences in number word systems can considerably alter number processing abilities in primary school children. For example, the number word systems in France and Wallonia (French speaking part of Belgium) differ insofar as in France the complex tens-structures of "soixante-dix" (literally: "sixty-ten", meaning 60+10) for 70 and "quatre-vingt-dix" (literally: "four-twenty-ten", meaning 4x20+10) for 90 are used, whereas the respective regular forms of "septante" and "nonante" are used in Wallonia. Seron and Fayol (1994) showed that due to these irregularities in the number word system, second graders from France made more errors on items comprising these numbers in different tasks (e.g., transcoding numbers from verbal to Arabic notation, transcoding numbers from verbal notation to representation with tokens, grammaticality judgements) compared to second graders from Wallonia.

Another study did not look at language effects on number processing in two groups of children from different countries, but between children speaking Welsh (regular number word system) or English (irregular number word system) both living in Wales (Dowker, Bala, & Lloyd, 2008). Dowker and colleagues showed that Welsh speaking children were better in magnitude comparison of two-digit numbers, but not in arithmetic.

3.1.2 Inversion effects on the development of numeracy

A specific irregularity in some number word systems (e.g., German, Dutch and Flemish, Danish, Maltese, Malagasy, Arabic, and partly in Czech and Norwegian; see Comrie, 2005) is the so-called inversion principle of two-digit number names (e.g., "drei-und-siebzig", literally "three-and-seventy", for 73). A detailed analysis of Austrian (German speaking) 7-year-old children's transcoding errors (Zuber et al., 2009) revealed that nearly 50% of the errors seemed to be related to this inconsistency of the German number word system, namely inversion. Another study (Helmreich et al., accepted) investigated the effect of inversion on first graders' number line estimation – a task that is frequently used to investigate children's number magnitude representation (e.g., Siegler & Opfer, 2003). In this number line estimation task, children are asked to mark on a horizontal line with the labelled end-points 0 and 100 where a specific number (e.g., 27) should go. Using this task, Helmreich and colleagues (accepted) compared the estimation accuracy of first graders from Austria (with inversion in the number word system) and Italy (without inversion in the number word system). They found that compared to Italian children, Austrian children were not only less accurate overall (unspecific inversion effect), but they were especially impaired for numbers with a larger distance between unit- and decade-digit (specific inversion effect). It is worth noting that both the study by Zuber and colleagues (2009) and by Helmreich and colleagues (accepted) investigated number processing abilities of first graders who have no formal experience with Arabic numbers larger than 20. Therefore, it is still unclear whether the reported inversion effects are restricted to multi-digit numbers children have not yet formally learned or whether these effects would also hold for older children. Yet, it is important to note that the children in the Zuber and the Helmreich studies were generally able to solve different tasks with stimuli from 0-100 quite well (Helmreich et al., accepted; Zuber et al., 2009).

In our study, we address the question whether specific and/or generalized linguistic (inversion) effects on the processing of Arabic numbers can also be observed in older (2nd grade) children who have already been formally taught all two-digit numbers.

3.1.3 Objectives of Study 2

The standardization of the dyscalculia test TEDI-MATH in three different languages (French, Flemish, and German) and five different regions/countries (France, Wallonia, Flanders, Austria, and Germany) allowed us to avoid the above mentioned difficulties by examining the effects of inversion on several tasks (writing multi-digit Arabic numbers, recognition of unit- and decade-digits, and subtraction) in large samples of children from multiple schools of five different countries/regions. With this approach, it is possible to explore whether differences in language properties have a general effect on numerical development or differentially affect specific numerical skills.

If the inversion property for double-digit numbers in German and Flemish (e.g. "dreiundsechzig" or "drieenzestig", literally "three-and-sixty") affects number processing abilities even in children who should already be able to transcode double-digit numbers, we expect an advantage in number processing of children speaking French (without inversion property for double-digit number; e.g. "soixante-trois", literally sixty-three) over children speaking German or Flemish in tasks using multi-digit numbers. Furthermore, we expect no differences between French and Walloon and no differences between Flemish, Austrian, and German children in the same tasks if the cross-cultural effects are really due to inversion. We do not explicitly expect an effect of inversion on subtraction of double-digit numbers, yet to our knowledge this possible transfer of inversion effects on calculation tasks has not been tested so far and is worth investigation.

Furthermore, we expect to better understand the cognitive processes underlying group differences by carrying out item analyses. Different performance patterns for specific item groups may help to differentiate in particular whether only specific effects of inversion (only for double-digit numbers) or also unspecific effects of inversion (also in decade-, teens-, or hundred-numbers) can be found.

3.2 Methods

3.2.1 Participants

Overall, 1957 children between kindergarten and third grade were tested for the standardization of the three TEDI-MATH versions [French: France and Wallonia (French speaking part of Belgium), Flemish: Flanders (Flemish speaking part of Belgium), and German: Germany and Austria].

In general, recruitment of the standardization samples took place in typical primary schools. Austrian children were (apart from a small subsample from Vienna) recruited from the rural area around Innsbruck and had mainly a middle-class socio-economic background. In Germany and Flanders, the participating schools were spread over the whole country and children came from mixed socio-economic backgrounds. In those samples, all children were asked to participate in order to represent the whole typical performance range. Children from Wallonia (French-speaking region of Belgium) and France were randomly selected in their respective country, but excluding children one or more years behind the regular grade corresponding to their age, or with intellectual or sensorial disabilities, or behavioural troubles.

In this study we will focus on the data of 220 children who were tested in the middle of second grade, as this age group showed the highest variability in performance. Results for the other age groups were similar, but less pronounced. Furthermore, in the middle of second grade all children are already expected to have mastered transcoding of double-digit numbers, which means that effects of inversion can not be attributed to incomplete knowledge about the number range up to 100.

Mean age of the 220 children was 91 months (sd = 5), ranging from 83 to 113 months. Sample characteristics of the five different groups can be obtained from Table 3.1.

French, Walloon, and Flemish children did not differ in their mean age (all pairwise $p > .05$, Bonferroni corrected) but were significantly younger than the German as well as the Austrian children (all $p < .05$, Bonferroni corrected). Therefore, age was used as a covariate in all ANOVAs.

Table 3.1: Sample characteristics: Number of children and age in months in each group (sd = Standard deviation)

	French	Walloon	Flemish	Austrian	German	Mean age (sd)
Boys	24	9	20	16	30	92 (6)
Girls	25	12	26	22	36	91 (5)
Total	49	21	46	38	66	91 (5)
Mean age (sd)	89 (2)	88 (2)	89 (6)	95 (5)	94 (4)	

3.2.2 Procedure and tasks

All children were assessed individually with the TEDI-MATH (Van Nieuwenhoven, Grégoire, & Noël, 2001) in a separate and quiet room in their schools by a trained tester. Written and informed consent was obtained from all parents or caregivers.

The TEDI-MATH is a multi-componential dyscalculia test based on cognitive neuropsychological models of number processing and calculation and has been tested for conceptual accuracy and clinical relevance in previous studies (e.g., Stock, Desoete, & Roeyers, 2009). Good psychometric properties were demonstrated (Kaufmann, Nuerk, Graf, Krinzinger, Delazer, & Willmes, 2009). For this study, we concentrated on three subtests with at least 10 items in the original French version (most subtests have more items in the Flemish and the German version).

In the Writing Arabic numbers to dictation-task, children were asked to write down 20 different Arabic numbers (3 single-digit, 9 double-digit, 8 three-digit numbers) that were presented verbally one by one by the tester.

For the Recognition of unit- and decade-digits-task, children were presented with Arabic multi-digit numbers on a sheet of paper and asked to point to the unit-digit or the decade-digit respectively (3 double-digit and 2 three-digit numbers each).

Subtraction was administered by showing the children one calculation problem after the other on a sheet of paper, asking them to solve them mentally as quickly and accurately as possible and to give the correct answer verbally.

All items were scored with 1 for a correct and 0 for an incorrect answer. All tasks were terminated after 5 consecutive errors. In this case, all other subsequent items were scored as incorrect. For all tasks, the particular items can be found in Tables A3.1-A3.3 in the Appendix.

3.2.3 Analyses

We conducted a univariate ANOVA with the factor group (French, Walloon, Flemish, German, and Austrian) and the covariate age in months on the number of correctly solved items for each of the four tasks.

To unambiguously test an effect of inversion, we used a family of tests to analyze this specific research hypothesis (see Westermann & Hager, 1986): We computed a one-way ANOVA to test a complex a-priori contrast on means from independent samples (Kohr & Games, 1977) between all children speaking a language with (Flemish, German, and Austrian children) vs. all children speaking a language without inversion (French and Walloon children). Furthermore, we conducted a two-sided independent samples t-test between the French and the Walloon as well as a univariate ANOVA between the Flemish, German, and Austrian children to analyze whether unexpected differences are found for groups that share a specific language characteristic (inversion yes or no). The specific research hypothesis can only be considered as being fully confirmed if the outcome is significant for the overall one-way ANOVA and not significant for the two-sided t-test and the specific ANOVA, respectively.

In case of warranted differentiation between specific and unspecific effects of inversion, contingency table tests for each item of a respective task were carried out.

3.3 Results

3.3.1 Writing Arabic numbers to dictation

The mean number of correctly solved items achieved in the task Writing Arabic numbers to dictation by each group (French, Walloon, Flemish, German, and Austrian) can be found in Figure 3.1.

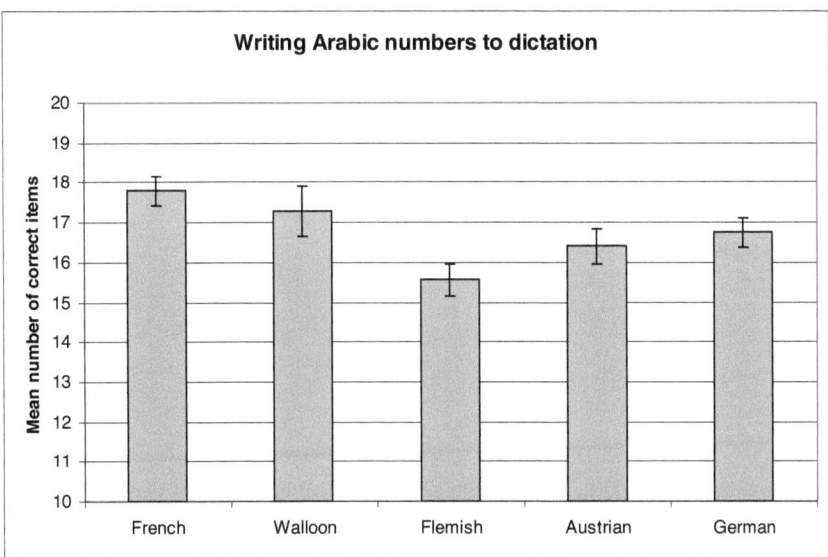

Figure 3.1: Mean number of correct items per group (French, Walloon, Flemish, Austrian, and German; error bars depict standard errors) for Writing Arabic numbers to dictation

A one-way ANOVA revealed a significant main effect of group [$F(4, 219) = 4.02$, $p = .004$], but no effect of age [$F(4, 219) = 0.03$, $p = .863$] for Writing Arabic numbers to dictation.

In the case of a direct effect of inversion on children's performance on this task, French and Walloon children (number word system without inversion) should be significantly better than Flemish, Austrian and German children (number word systems with inversion), but no significant differences should be found between the groups who share the (non)existence of this linguistic property (see above for analyzing substantive hypothesis by using a family of tests). All these three assumptions could be corroborated, because the a-priori contrast between French and Walloon vs. Flemish, Austrian, and German children was significant [$t(1, 215) = 15.42$, $p < .001$], and the two-sample t-test between French and Walloon children [$t(1, 68) = .72$, $p = .473$] as well as the ANOVA between Flemish, Austrian, and German children [$F(2, 149) = 2.36$, $p = .098$] were not significant.

To analyse whether this linguistic effect was due to poorer performance of children speaking a language with inversion in the number word system only on items with double-digit numbers (specific inversion effect) or on numbers without decade-unit-structure as well (unspecific inversion effect), we carried out contingency table tests on each item (see Table A3.1 in the Appendix).

The following items revealed significant group differences with more French and Walloon children than expected and less or as many of the other three groups than expected being correct, therefore showing an inversion effect also at the item level: item "68" (p = .009), "150" (p < .001), "101" (p = .024), "643" (p = .004), "190" (p = .050), and "951" (p = .006).

3.3.2 Recognition of unit- and decade-digits

The mean number of correctly solved items achieved in the task Recognition of unit- and decade digits by each group (French, Walloon, Flemish, German, and Austrian) can be obtained from Figure 3.2.

A one-way ANOVA revealed a significant main effect of group [$F(4, 219) = 25.27$, $p < .001$], but no effect of age [$F(4, 219) = 0.62$, $p = .430$] for Recognition of unit- and decade-digits.

We tested for the effect of inversion on Recognition of unit- and decade digits by conducting the same family of tests as for the task Writing Arabic numbers to dictation (see 3.2.3). The contrast between the performance of children speaking a language with vs. without inversion was significant [$t(1, 39.5) = 6.37$, $p < .001$; equal variances not assumed], which is in line with the hypothesis of an inversion effect. Yet, the two-sample t-test between French and Walloon children [$t(1, 26.6) = 4.95$, $p < .001$; French children better; equal variances not assumed], as well as the ANOVA between Flemish, Austrian, and German children [$F(2, 149) = 23.73$, $p < .001$] were significant as well, which does not support the notion of a purely linguistic effect for this task. Post-hoc tests showed that within the group of children speaking a language with inversion, Flemish children were worse than Austrian and German children (both $p < .001$, Bonferroni-corrected), but Austrian and German children did not differ in their performance ($p > .99$, Bonferroni-corrected).

To better understand the cause for group differences in this task, we conducted contingency table tests in analysing specific items.

The counts and expected counts for the five groups, X^2-test-value, and level of significance for each item of Recognition of unit- and decade-digits can be obtained from Table A3.2 in the Appendix.

The chi-square test was highly significant for all items, indicating group differences in item difficulty (all $p < .001$). Descriptively, the pattern of results was the same for all items:

many more French children than expected were correct, more German and Austrian children than expected were correct for most items (German children: 10 out of 10; Austrian children: 8 out of 10), and on all items less Walloon as well as Flemish children than expected were correct.

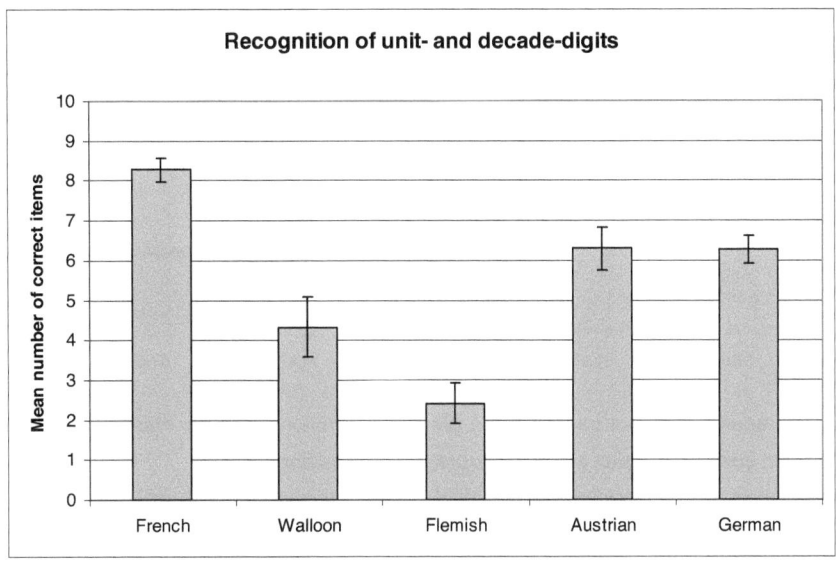

Figure 3.2: Mean number of correct items per group (French, Walloon, Flemish, Austrian, and German; error bars depict standard errors) for Recognition of unit- and decade digits

3.3.3 Subtraction

The mean number of correctly solved items achieved by each group in the task Subtraction (French, Walloon, Flemish, German, and Austrian) is depicted in Figure 3.3. A one-way ANOVA revealed a significant main effect of group [$F(4, 219) = 4.76$, $p = .001$], but no effect of age [$F(4, 219) = 0.27$, $p = .613$] for Subtraction.

As we did not have any specific hypotheses about group differences on this task, we carried out post-hoc tests for group comparisons. They revealed that French children scored lower than children from Austria and Germany (all $p < .001$, Bonferroni-corrected), but no other significant group differences emerged (all other $p > .10$).

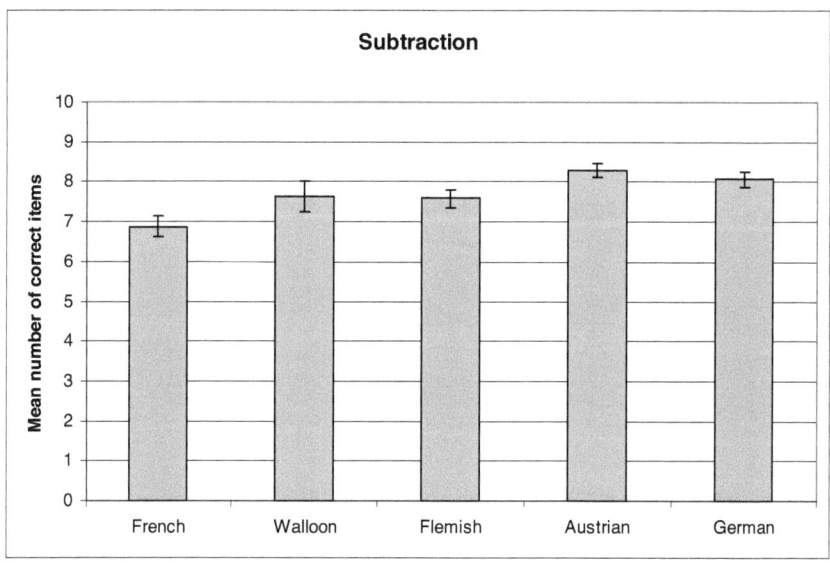

Figure 3.3: Mean number of correct items per group (French, Walloon, Flemish, Austrian, and German; error bars depict standard errors) for Subtraction

As no known national curricular differences would have led us to expect lower performance of French children in Subtraction, we again conducted contingency table tests for all items, as an in-depth evaluation of a potential curricular effect is warranted (see Table A3.3 in the Appendix for counts and expected counts of the five groups, X^2-test value, and levels of significance for each item of Subtraction).

The following items showed the same pattern, with only the French children being worse than expected and at least as many children as expected being correct in all other groups: "9-5" ($p = .003$), "5-3" ($p = .008$), "6-6" ($p < .001$), and "40-20" ($p = .001$). There were also two items with a (marginally significant) different pattern of group performance: both "16-4" ($p = .058$) and "27-6" ($p = .053$) were correctly solved by less French as well as Walloon children than expected and by at least as many as expected Flemish, German, and Austrian children.

3.4 Discussion

In this study, we found inversion effects on one out of three mathematical skills of second graders by comparing the standardization samples of the dyscalculia test TEDI-MATH (Van Nieuwenhoven et al., 2001; Grégoire, Noël, & Van Nieuwenhoven, 2004; Kaufmann et al., 2009) from France, Wallonia, Flanders, Austria, and Germany.

In summary, the hypothesis of an inversion effect on multi-digit number processing could be clearly confirmed for Writing Arabic numbers to dictation, but not for Recognition of unit- and decade-digits, both at the task level as well as by detailed item analyses.

Unexpected patterns of group differences were found for the task Recognition of unit- and decade-digits as well as for Subtraction. In the following, we will first discuss our more specific research questions concerning inversion effects and then these unexpected performance patterns.

3.4.1 Inversion effects between language groups

Two recent studies support the notion that the inversion principle of two-digit numbers found in some languages (e.g., naming 23 literally as "three-and-twenty" like in German and Dutch) is a irregularity in a number word system that may impair the development of number processing skills in children (see Zuber et al., 2009; Helmreich et al., accepted). Yet, both these studies investigated the number processing skills of first graders who are not yet formally taught double-digit Arabic numbers above 20. Therefore, it was not clear whether the reported inversion effect could also be found in older children who were already formally taught the Arabic number system up to 100. Our results unambiguously showed that French speaking children were significantly better in writing multi-digit numbers than children whose mother tongue features the inversion principle for two-digit number names, even at an age when children are already expected to master two-digit Arabic numbers according to the curriculum.

3.4.2 Specific vs. unspecific effects of inversion

Furthermore, specific as well as unspecific effects of inversion on writing multi-digit numbers have been found: The group differences for the items "68", "643", and "951" can be easily explained by a specific inversion effect on the processing of double-digit numbers, whereas the significant differences for the items "150", "101", and "190" suggest that the inversion property of a number word system may also exert an unspecific negative influence on the transcoding of multi-digit numbers.

This finding seems to be in line with the studies by Helmreich and colleagues (accepted) and by Zuber and colleagues (2009) who found both specific and unspecific effects of inversion on the number processing abilities of Austrian first graders. The latter study interpreted part of Austrian first grader's transcoding errors for three-digit numbers to be an overgeneralization of the inversion principle – that is, "Write the first spoken digit on the rightmost position" (e.g., writing the dictated number "400" as "104"). Interestingly, our cross-cultural comparisons can not confirm this particular interpretation of the error analyses by Zuber and colleagues (2009), as the non-significant group differences for the hundreds-numbers "200" and "700" in our study are not in line with an effect of inversion on this type of numbers. The outcomes of the item analyses for these two items are not only not significant ($p = .159$ and $p = .119$, respectively), but the descriptive disparities between counts and expected counts clearly do not follow a pattern as predicted by an effect of inversion: Walloon children score lower and German and Austrian children higher than expected on "200", and Austrian children higher than expected on "700". Furthermore, an overgeneralization of the inversion principle for three-digit numbers would not explain why an effect of inversion was also found for our item "101".

Hence, we propose that this error type as proposed by Zuber and colleagues (2009) might be due to a general and language-independent inappropriate application of the multiplicative principle of multi-digit Arabic numbers (see ADAPT model by Barrouillet et al., 2004) and not due to the inversion principle in the German and Flemish number word systems. Alternatively, we assume that this irregularity in a number word system constitutes a general obstacle for understanding the regularity of the base-10 notational system (or place-value system) of Arabic numbers, and not only one more syntactical rule children have to acquire during the development of transcoding (but see Barrouillet et al, 2004, for a different view). This interpretation is in line with both the study by Helmreich and colleagues (accepted) who found an inversion effect on the number line estimation task thought to picture children's number magnitude representation, and with cross-cultural comparisons between Asian and English speaking children, which consistently show better general place-value understanding in the first group, namely the children speaking the language with the more transparent and regular number word system (for an overview: see Geary et al., 1996). Yet, another possible explanation accounting for both our results and the overgeneralization of the inversion rule as reported by Zuber and colleagues (2009) would be that the children in our study were half a year older than the ones tested by Zuber and colleagues. Therefore, it may be possible that an overgeneralization of the inversion rule for transcoding three-digit numbers can only be found at a very specific time window, namely during first grade.

3.4.3 Positive effects of inversion on Subtraction

Surprisingly, we found a positive (i.e. error-reducing) effect of inversion on two items of Subtraction, namely "16-4" and "27-6". These two items have in common that a single-digit number has to be subtracted from a two-digit number. Therefore, we speculate that naming the unit-digit first in the minuend may enhance attention to the digit from which the second number should be subtracted, at least during a short time window in children's numerical development. This may be interpreted as a positive effect of the inversion property in a number word system for learning how to subtract.

3.4.4 No inversion effect on Recognition of unit- and decade-digits

Concerning the unexpected performance pattern we found for the task Recognition of unit- and decade-digits, the group differences can not be clearly interpreted as being indicative of an inversion effect. We did not find the expected advantage of all French speaking children over the others, as Walloon children scored not only lower than expected but even worse than German and Austrian children. If inversion would impair performance on this task, we would expect Walloon children to be better than German speaking children. Furthermore, teen numbers and simple decade numbers were also affected by the group factor, which is not expected if performance is influenced by inversion.

Alternatively, these results may suggest differences in the focus of math curricula in the different countries, with the strongest emphasis on understanding the value of digits in multi-digit numbers in France, followed by Germany and Austria, and the lowest in Belgium (Wallonia and Flanders). To the best of our knowledge, national math curricula have largely comparable study aims. Thus, we can only speculate about this issue. Yet, two other findings of our study support this interpretation:[2] First, French children are the best in the other task requiring multi-digit number processing, namely Writing Arabic numbers to dictation. Second, French children are the worst in Subtraction. A possible explanation unifying both findings is that in France math curricula for first and second graders set priorities more on multi-digit number processing and less on calculation compared to Belgium, Austria, and Germany.

The presumably curricular effect on the task Recognition of unit- and decade-digits does not explain why inversion should be less important for this task compared to Writing multi-digit numbers to dictation. One possible explanation may be that our task used to assess recognition of unit- and decade-digits does not require the processing of number words and is therefore not affected by irregularities of a number word system. Another possibility may be that inversion negatively affects understanding of the regularity of the base-10 notational system of Arabic numbers (see above), and that this understanding is more important for writing Arabic numbers than for recognizing unit- and decade-digits for two reasons: On the one hand, children have to actively produce multi-digit numbers in the first task, which can only be done correctly with an adequate base-10 understanding. On the other hand, correctly pointing to the unit- or the decade-digit may be possible by applying just two simple explicit rules like "The unit-digit is always the rightmost one" or "The decade-digit is always the second rightmost one" without really understanding the base-10 system of Arabic numbers which defines what the unit- and the decade-digits mean. Differential effects of inversion on these two tasks may thus be due to task requirements.

---

[2] Additional support for the interpretation of curricular effects on this task comes from the standardization of the German TEDI-MATH Version (Kaufmann et al., 2009), as this task was the only one in which effects of individual class and/or teacher could be observed. The respective analysis was only carried out for second grade second semester, as only in this age group more than 9 children of one single class were tested in more than two instances. A Kruskal-Wallis-test revealed a significant performance difference in this age group between children of different classes for Recognition of unit- and decade-digits ($p < .001$), but not for Writing Arabic numbers to dictation ($p = .064$), Subtraction ($p = .211$), or Multiplication ($p = .119$).

3.5 Conclusion

In general, the results of our study show that numerical skills do not develop in a unitary fashion. Rather, individual performance on specific mathematical skills can be differentially influenced by linguistic properties of the number word system. This means that for a promising investigation of cross-cultural differences in mathematical development, it seems fruitful to study the performance patterns of children from multiple countries, as it is very likely that cultural factors are confounded if only two countries are compared. For example, if we would have examined only German and French children, we would have misinterpreted all group differences as linguistic effects on numerical achievement of these two groups, which could be ruled out by specific item analyses. Furthermore, our results showed that two tasks tapping multi-digit number processing (Recognition of unit- and decade digits and Writing of Arabic numbers to dictation) were differentially affected by the existence of the inversion property in the respective number word systems. This calls for a multi-task approach in cross-cultural comparisons, but in many instances only one or very few tasks have been used so far. Our results show that cross-cultural differences in one task are not readily transferable to another task. What is more, sometimes even such a between-task approach may be too unspecific; rather, item effects within tasks have to be investigated. In some cases in our study, the effects of interest only came out clearly for subsets of items and not at the task level. In our view, this conspicuously shows that a profound understanding of processes underlying the development of numerical skills is only possible with the use of differentiated measures.

4 Study 3: Influence of gender equality on gender differences in numerical tasks

4.1 Introduction

Study 3 was carried out to examine whether the gender differences we found in Study 1 in the German standardization sample of the TEDI-MATH (Kaufmann et al., 2009) are related to a socio-cultural measure of gender equality. We observed unexpected and profound performance differences between boys and girls in several numerical tasks (see 2.3.1). The male advantage was stronger for subtests tapping on the component multi-digit number processing, but found in calculation subtests as well.

In general, gender differences have been frequently reported for mathematical performance (e.g., Mills, Abland, & Stumpf, 1993; Robinson, Abbott, Berninger, & Busse, 1996, Robinson, Abbott, Berninger, Busse, & Mukhopadhyay, 1997; TIMMS III, 1996; TIMMS 2007) and especially for mathematical problem solving (Kimura, 2000). Yet, very little is known about the development and the causes of these differences.

One of the first and most influential meta-analyses studying this topic revealed that gender differences in mathematical problem solving emerge only at the high-school level (Hyde, Fennema, & Lamon, 1990). Therefore, gender differences in mathematical cognition are not expected to be observed in primary school children. For example, in a study at regular elementary and secondary schools in Belgium, a large overlap was found between the different mathematical performances of girls and boys (Desoete, 2007; 2008). There was a limited impact of gender differences on mathematical problem solving in primary school (total sample n = 2.255). However, males did better in almost all mathematics tasks at the end of secondary school in Flanders (n = 796).

4.1.1 Higher performance variability in males as a possible reason for gender differences in mathematics

One general possible explanation for a male advantage in studies about mathematical achievement was provided by Hedges and Nowell (1995). In an analysis of six studies from the USA about different cognitive tasks that used national probability samples they showed that the test scores of males generally have larger variance and that therefore males typically outnumber females among high-scoring groups. However, an analysis of the PISA 2003 mathematics assessment in the United States (Program for International Student Assessment; OECD, 2003) by Liu, Wilson, and Paek (2008) yielded a different result. In this sample of over 5.000 15-year-old students, gender differences were overall small, but performance distributions of boys and girls differed according to the four mathematical domains tested. The domain Space and Shape (e.g., mental rotation of objects) displayed the largest gender gap (effect size $d = .14$), with more girls in the lower and more boys in the higher performance ranges. In the domain Change and Relationship (e.g., translation between representation formats, understanding of fundamental relationships and types of change) girls were overrepresented in the middle performance range and underrepresented in the higher performance range ($d = .10$). No significant gender differences were found for the mean performance ($d = .04$) or the performance distribution in the domain Quantity (mostly computational items). The domain Uncertainty (e.g., calculation of probabilities) showed the second highest overall gender difference ($d = .12$) due to more females belonging to relatively low performing groups and more males belonging to the middle performing groups.

Therefore, gender differences seem to be not only due to higher performance variability among males, and are not found in all numerical domains.

This finding of differential gender differences on various mathematical skills indicates that using different tasks and not one overall score for mathematical ability may be more promising in revealing possible causes for respective gender differences as compared to global math performance measures.

4.1.2 Socio-cultural reasons for gender differences in mathematics

Concerning these possible causes for performance differences between boys and girls in mathematics, the possibility most relevant for our society is that socio-cultural factors may underlie the male advantage in mathematical reasoning. One hint for this explanation is that in general gender differences have been found to be smaller in studies conducted after 1974 compared to older studies (Hyde, Fennema, & Lamon, 1990). Several studies have focused on possible mediators of culturally determined gender differences in mathematics and found, e.g., that lower parental expectations of females' math achievement influenced their attitudes towards math (Eccles & Jacobs, 1986), that the coupling of self-believes about mathematical abilities, actual aptitude, and attitudes towards mathematics (Denissen, Zarrett, & Eccles, 2007) increase over time, that math anxiety was already higher in females than in males in elementary school (Krinzinger, Kaufmann, Dowker, Thomas, Graf, Nuerk, & Willmes, 2007), and that stereotype threats were indeed a significant predictor for individuals' math performance scores (Osborne, 2001).

Evidence for cultural mediation of the gender gap in math comes from recent studies that found significant correlations between national measures of gender equality like the World Economic Forum's Gender Gap Index (GGI; Hausmann, Tyson, & Zahidi, 2006) and presence of females in national teams for the International Mathematic Olympiad (IMO; www.imo-official.org) as presented by Hyde and Mertz (2009), or a study by Guiso and colleagues (2008), who compared the results of over 246.000 15-year-old boys and girls from over 40 countries having taken part in the 2003 PISA study (OECD, 2003) and correlated the results with the GGI. The most important result of this study was that "the gender gap in math scores disappears in countries with a more gender-equal culture" (Guiso et al, 2008, p. 1164; for a similar conclusion see Else-Quest, Hyde, & Linn, 2010).

4.1.3 Fact retrieval advantage as a possible reason for gender differences in mathematics

A completely different account for better scores of males in standardized math achievement tests was suggested by Royer, Tronsky, Chan, Jackson, and Marchant (1999). In college students, they found a male advantage in math fact retrieval and concluded that higher computational fluency should leave more time and cognitive resources for solving more complex tasks in math tests, especially under time pressure. Testing both a spatial cognition hypothesis and the math fact retrieval hypothesis directly in a large sample of college students, Geary, Bow Thomas, Liu, and Siegler (2000) found that both spatial cognition and computational fluency were mediating the male advantage in arithmetical reasoning.

4.1.4 Objectives of Study 3

Despite these ongoing debates about gender differences in mathematical abilities and the scientific efforts to unveil their possible causes, surprisingly little is known so far about their developmental origins. Only very rarely gender differences in math performance have yet been reported for primary school children (e.g. Fennema et al., 1998; Rosselli et al., 2008; but see 2.3.1). Furthermore, it is even less clear to which extent cultural factors like gender equality influence the development of a possible male advantage in some numerical skills.

In our study we wanted to uncover (i) whether gender differences we found in numerical tasks in a standardization sample from Austria and Germany for the dyscalculia testbattery TEDI-MATH (Kaufmann et al., 2009: see 2.3.1) can be replicated in a French and a Belgian sample, and if so, (ii) whether respective gender differences may be already socio-culturally mediated even in primary school children.

Concerning possible gender differences in our samples, socio-cultural mediation should result in a differential pattern following the Gender Gap Index (GGI; Hausmann et al., 2006), which was also used in the study by Guiso and colleagues (2008). If gender differences in mathematical abilities of primary school children are already influenced by socio-cultural factors, they should be most pronounced in French and least pronounced in German children (see 4.2.2).

4.2 Method

4.2.1 Participants and tasks

The same sample and tasks (plus the subtest Multiplication with the items 1x7, 6x1, 2x4, 3x3, 8x0, 3x5, 4x4, 10x2, 0x3, and 3x10) were used as for Study 2 (see 3.2.1 and 3.2.2).

4.2.2 Gender Gap Index

The Gender Gap Index (GGI; Hausmann et al., 2006) focuses on measuring gaps rather than levels, captures gaps in outcome variables rather than gaps in means or input variables, and ranks countries according to gender equality rather than women's empowerment. The four subindices it is composed of are: Economic participation and opportunity, educational attainment, health and survival, and political empowerment. The GGI is then calculated for 128 nations/countries, and their respective rank pictures whether the gender gap is small (low ranks) or large (high ranks) in comparison to the other countries. Of the four countries included in our study, Germany showed the smallest GGI (rank 5 of 128), followed by Austria (rank 26 of 128), Belgium (rank 33 of 128), and France (rank 70 of 128; see Hausmann et al., 2006).

4.2.3 Analysis

We conducted several 2x4 ANOVA's with the factors gender (male, female) and country (Germany, Austria, Belgium, France) on the number of correctly solved items for each of the four tasks. Differences between countries were the topic of Study 2 and will therefore not be discussed here as no significant interactions were observed.

For tasks with significant main effects of gender, we correlated national ranks of the gender gap indices (GGI; Hausmann et al., 2006) with the respective mean performance differences between boys and girls in order to investigate a possible socio-cultural influence on the gender differences in specific numerical skills.

4.3 Results

4.3.1 Writing Arabic numbers to dictation

The mean number of correctly solved items achieved in the task Writing Arabic numbers to dictation by each group (Germany, Austria, Belgium, and France) separately for girls and boys can be found in Figure 4.1.

A 2 (gender) x 4 (country) ANOVA revealed both a significant main effect of gender in favor of boys [$F(1, 219) = 34.97$, $p < .001$] and a significant effect of country [$F(3, 219) = 3.69$, $p = .013$], but no significant interaction [$F(3, 219) = 1.74$, $p = .16$] for Writing Arabic numbers to dictation.

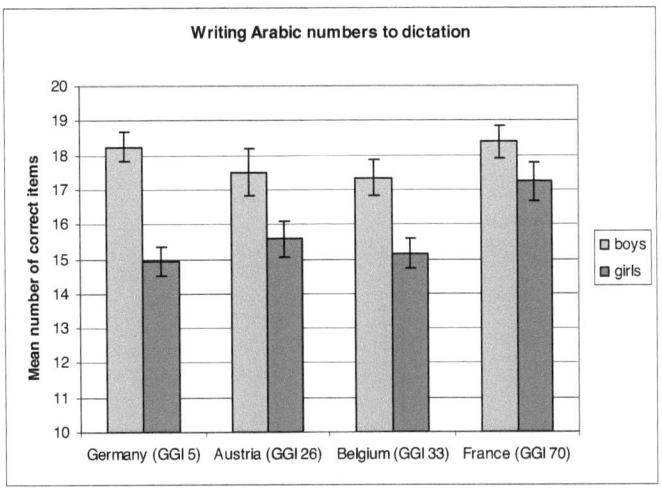

Figure 4.1: Mean number of correct items in Writing Arabic numbers to dictation by country (Germany, Austria, Belgium, and France) separately for boys and girls (error bars depict standard errors)

If the main effect of gender on the task Writing Arabic numbers to dictation were due to the socio-cultural factor of gender equality as measured by the GGI (Hausmann et al., 2006), the performance differences between boys and girls should be the largest in France, second largest in Belgium, even smaller in Austria, and the smallest gender difference being found in Germany. Yet, the mean differences between correctly solved items of boys and girls per country clearly do not follow this pattern: The difference was largest in Germany (mean difference: 3.3), followed by Belgium (2.2), Austria (1.9), and the smallest in France (1.1). Moreover, the correlation between the national gender differences in performance and the respective country's GGIs was negative (Spearman's rho = -.80, p = .10). In general, a test on significance based on a sample of four is not very informative, and the size of the correlation coefficient should be taken into account.

4.3.2 Recognition of unit- and decade-digits

The mean number of correctly solved items achieved in the task Recognition of unit- and decade digits by each group (Germany, Austria, Belgium, and France) separately for girls and boys can be obtained from Figure 4.2.

A 2 (gender) x 4 (country) ANOVA revealed no significant main effect of gender [$F(1, 219) = 2.17$, $p = .140$], a significant effect of country [$F(3, 219) = 30.57$, $p < .001$], and no significant interaction [$F(3, 219) = 0.48$, $p = .70$] for Recognition of unit- and decade-digits. Therefore, no further analyses were conducted.

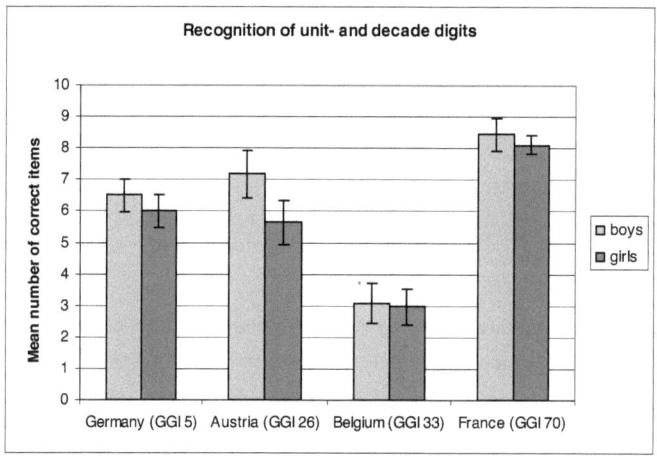

Figure 4.2: Mean number of correct items in Recognition of unit- and decade-digits by country (Germany, Austria, Belgium, and France) separately for boys and girls (error bars depict standard errors)

4.3.3 Subtraction

The mean number of correctly solved items achieved in the task Subtraction by each group (Germany, Austria, Belgium, and France) separately for girls and boys is depicted in Figure 4.3.

A 2 (gender) x 4 (country) ANOVA revealed a significant main effect of gender in favour of boys [$F(1, 219) = 10.55$, $p = .001$] and a significant effect of country [$F(3, 219) = 7.36$, $p < .001$], but no significant interaction [$F(3, 219) = 0.54$, $p = .659$] for Subtraction.

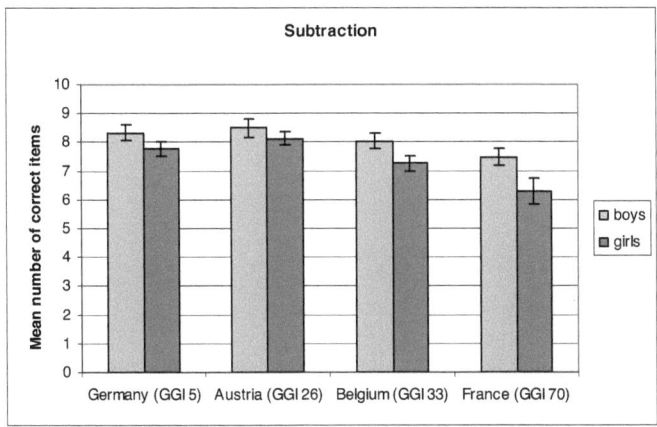

Figure 4.3: Mean number of correct items in Subtraction by country (Germany, Austria, Belgium, and France) separately for boys and girls (error bars depict standard errors)

Again, we checked the national mean performance differences between boys and girls against the respective gender gap indices and found that the group patterns were almost corresponding (GGI ranks: Germany = 5, Austria = 26, Belgium = 33, France = 70; mean performance differences: Austria = 0.4, Germany = 0.6, Belgium = 0.8, France = 1.2). This descriptive observation was confirmed by a positive correlation between the national GGIs and the respective mean performances differences between boys and girls (Spearman's rho = .80, p = .10).

4.3.4 Multiplication

The mean number of correctly solved items achieved in the task Multiplication by each group (Germany, Austria, Belgium, and France) separately for girls and boys are shown in Figure 4.4.

A 2 (gender) x 4 (country) ANOVA revealed a significant main effect of gender in favour of boys [$F(1, 219) = 17.99$, $p < .001$], and a significant effect of country [$F(3, 219) = 27.38$, $p < .001$], but no significant interaction [$F(3, 219) = 0.68$, $p = .567$] for Multiplication.

As for all other tasks with a significant main effect of gender, we analyzed the correlation between the gender differences in Multiplication of all groups (mean performance difference per group: Belgium = 1.0, Austria = 1.3, Germany 1.8, France = 2.4) and the respective national GGIs, which was not significant (Spearman's rho = .20, p = .40).

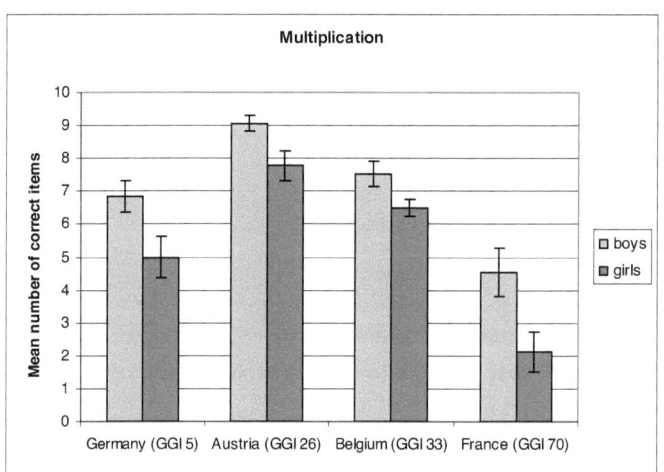

Figure 4.4: Mean number of correct items in Multiplication by country (Germany, Austria, Belgium, and France) separately for boys and girls (error bars depict standard errors)

4.4 Discussion

The two main goals of this study were the following: (1) Find out whether gender differences in favour of boys we had found in various math tasks for German speaking children could be replicated for French and Belgian children, and if so, (2) examine whether respective differences in different numerical tasks are related to the national Gender Gap Indices (GGI; Hausmann et al., 2006) as a recent study by Giuso and colleagues (2008) showed for 15-year-olds.

The significant main effect of gender in favor of boys on three out of four tasks (Writing Arabic numbers to dictation, Subtraction, and Multiplication) in all four samples of young primary school children was not necessarily anticipated (Hyde & Mertz, 2009) and constitutes an interesting finding in itself.

Furthermore, we found a profound correlation between the national GGIs and the performance differences between boys and girls in the four groups for Subtraction, but not for Multiplication. For Writing Arabic numbers to dictation it was as strong as for Subtraction, but negative. This may be seen as evidence for a very early influence of the socio-cultural factor of gender equality on the gender differences in some, but not all specific numerical skills. To this end, the male advantage in some mathematical tasks (especially writing multi-digit numbers) can not be fully explained by socio-cultural factors.

In this context it is important to note that in all three tasks with significant gender differences the most difficult items were ahead of the respective national math curricula and that children were not expected to solve them correctly. In other words, the performance differences between boys and girls were attributable to skills they have not yet been explicitly taught. This is in line with the frequent finding of a male advantage in standardized mathematical problem solving tasks, but not for school achievement tests in mathematics (Kimura, 1999). Yet, this observation does not answer at all the question why boys are better in some mathematical tasks even at that young age if this male advantage is not related to socio-cultural factors.

Alternative explanations of why male subjects outperform females on at least some mathematical skills come from more cognitive perspectives, sometimes with an emphasis on biological causes (for an overview, see Geary, 1996). An influential line of research claims that males have better visual-spatial abilities or rely more strongly on spatial problem solving strategies compared to females, which provides them with an advantage for mathematical problem solving (Casey et al., 1995, 2001; Fennema et al., 1998; Rosselli et al., 2008) or numerical representations (Lonnemann et al., 2008). This possibility will be further investigated in the next study.

4.5 *Conclusion*

Gender differences in subtraction seem to be mediated by socio-cultural factors like national gender equality even in young children, whereas this explanation did not hold for the male advantage in writing multi-digit numbers in the same children. Therefore, alternative explanations for performance differences in boys and girls in some numerical tasks need to be found.

5 Study 4: What accounts for gender differences in multi-digit number processing?

5.1 Introduction

In Study 3 we showed that the significant gender differences in favour of boys which we observed for the transcoding task Writing Arabic numbers to dictation (see also Zuber et al., 2009) in four different countries were not related to a measure of the socio-cultural gender gap (see 4.3.1), namely the GGI (Hausmann et al., 2006), in opposition to the gender differences in the task Subtraction which were related to the GGI (see 4.3.3).

5.1.1 The spatial cognition hypothesis of gender differences in mathematical cognition

We speculated that gender differences in transcoding multi-digit numbers may be more due to differences in general visual-spatial abilities (Casey et al., 1995, 2001; Fennema et al., 1998; Rosselli et al., 2008) than to socio-cultural factors. Most prominently, Geary (1996) argued that sexual selection during human phylogeny and associated mechanisms like sex hormones resulted in a male advantage in spatial processing. A meta-analysis by Linn and Petersen (1985) found pronounced performance differences between males and females in spatial perception ($d = .44$; 81 studies) and mental rotation ($d = .94$; 18 studies). This male advantage may cause boys to be more interested in spatial objects and numbers than girls, and to use spatial solution strategies for mathematical problem solving more often (Geary, 1996). This spatial cognition hypothesis was supported by several studies conducted by Casey and her co-workers. They showed that gender differences in the Scholastic Aptitude Test-Math (SAT-M) were mediated by mental rotation ability in a large sample of young adults (Casey, Nuttal, & Benbow, 1995). They also observed that mental rotation ability in a sample of female college students was accounted for by an interaction of familial handedness patterns as indicator for biologically based individual differences and choice of college major as indicator for environmental factors (Casey, 1996). Finally, in a large sample of 8-graders, they showed that lower math scores of females were completely explained by their relatively poorer spatial-mechanical skills (74% of total indirect effects) and their lower mathematics self-confidence (26% of total indirect effects; Casey, Nuttall, & Pezaris, 2001). Rosselli and colleagues (2008) also found that spatial abilities were mediating gender effects in mental mathematical operations and in solving arithmetical problems in children from age 7 onwards. In line with Geary (1996), Casey and her co-workers (2001) interpreted their own findings to be due to different problem solving styles of males and females, with the latter being less likely to use spatial strategies that are more effective than algorithmic strategies for mathematical problem solving (van Garderen, 2006). This hypothesis was confirmed by directly investigating mathematical problem solving strategies of boys and girls in a sample of high-school students (Gallagher, De Lisi, Holst, Gillicuddy-De Lisi, Morely, & Cahalan, 2000), by a longitudinal study in primary school children from 1st to 3rd grade (Fennema, Carpenter, Jakobs, Franke, & Lewi, 1998), and by the observation that females displayed larger performance increases relative to males after being explicitly trained in spatial representation of mathematical relationships (Johnson, 1984). Furthermore, differential relations between spatial numerical representations and calculation skills in male vs. female third graders have been reported (Lonnemann, Krinzinger, Knops, & Willmes, 2008).

Finally, a study by Zuber and colleagues (2009) indicated that individual differences in visual-spatial working memory capacity may be another possible cause for performance differences in multi-digit number processing, as they could show that visual-spatial working memory, but not phonological working memory capacity, was a significant predictor for the number of transcoding errors in first graders (see Raghubar et al., 2010, for a general overview about the relevance of working memory for mathematics).

5.1.2 The psychobiosocial model of gender differences in mathematics

Yet, attitudes towards mathematics have also been shown to be a possible mediator of culturally determined gender differences in mathematics (Eccles & Jacobs, 1986). Moreover, the coupling of self-believes about mathematical abilities, actual aptitude, and attitudes towards mathematics was shown to increase within adolescence (e.g., Denissen, Zarrett, & Eccles, 2007). It was also shown that even in primary school evaluation of mathematics was higher in boys compared to girls (Krinzinger et al., 2007), and that calculation ability and evaluation of mathematics influenced each other reciprocally at the beginning of primary school (Krinzinger, Kaufmann, & Willmes, 2009).

Taken together, the spatial cognition hypothesis and the role of attitudes on gender differences in mathematics can be seen as a good example of the psychobiosocial model proposed by Halpern (1997), namely that biological predispositions, individual interests, social environmental factors, and gender differences in mathematics are interacting during a cognitive learning process.

5.1.3 Objectives of Study 4

If visual-spatial abilities, visual-spatial working memory capacity, and evaluation of mathematics can account for gender differences in Writing Arabic numbers to dictation, this should also hold for the observed gender differences in other tasks using multi-digit numbers (see 2.3.1), as they were found to load on one cognitive component (multi-digit number processing as compared to a calculation component; see 2.4.2).

Therefore, we wanted to investigate the developmental relations between sex, evaluation of mathematics, visual-spatial working memory, visual-spatial abilities, and multi-digit number processing in early primary school years. We did this by using a longitudinal approach in a new sample using slightly different tasks as in the studies presented before.

5.2 Methods

5.2.1 Participants

We tested n = 149 children attending normal schooling at five different schools supporting mostly middle-income families in Aachen, Germany. Written and informed consent was obtained by all parents, teachers, and headmasters involved. Seven children of the initial sample had to repeat a school year due to severe learning problems (which is a quite frequent procedure in the German educational system) and were therefore excluded from the sample. Another two children were excluded for withdrawn consent (once by the parents, and one child refused to participate after the second testing session), resulting in a final sample of n = 140 children. Of these 140 children, 80 (57.1%) were girls and 60 (42.9%) were boys. Almost all children in the sample were Caucasian (96%). Mean age at the end of first grade when children first participated in our study was seven years and six months (SD = 4 months), ranging from six years eight months to eight years five months. Five children moved away during the overall testing period of one and a half years. As none of them had shown any learning problems as long as they had participated in the study, their missing values (< 5%) were treated like values missing-at-random and were replaced by the means of all other children at the level of task raw scores. Participation in the experiment was voluntary and was rewarded by a colourful pen as a small incentive.

5.2.2 General procedure

The same children were tested at three different time points, with approximately half a year between the testing sessions (T1: end of first grade; T2: middle of second grade; T3: end of second grade). Children were presented with two transcoding tasks and an Arabic magnitude comparison task to assess children's multi-digit number processing performance, a self-rating scale concerning their evaluation of mathematics, two subtests of the Beery-Buktenica Developmental Test of Visual-Motor Integration (VMI; Beery, 1997) to investigate their general visual-spatial abilities, and the Corsi-Block-Tapping task (Corsi, 1972) forwards and backwards for their visual-spatial working memory capacity.
The Reading Arabic numbers task, the Magnitude comparison task, and the tasks tapping visual-spatial working memory capacity were always administered in individual testing sessions, as well as the writing Arabic numbers task, the self-rating questionnaire, and the visual-spatial tasks at T1 and T2. The latter tasks were assessed in groups of 15-20 children each at T3.

5.2.3 Stimuli and task procedures

Reading, writing, and comparing Arabic numbers task
For the Reading Arabic numbers task (RN), children were asked to read aloud different Arabic numerals in pseudo-randomized order. At T1, children were presented with more double-digit and less four-digit numbers compared to the following three time points.

Overall time for completing this task was measured in seconds. For the Writing Arabic numbers task (WN), children were asked to write down different numbers in Arabic format on dictation (up to four-digit numbers; more double-digit and less four-digit numbers at T1). As this task was administered in group testing sessions at T3, we could not measure overall processing time per child. The stimulus sets for RN and WN were composed according to syntactic features of the numbers (two items of all possible combinations of two-, three- and four-digit numbers with the numbers from "one" to "nine" on the leftmost position and using single-digit numbers, teens, decades, double-digit numbers, and zeros for the digits in the positions with smaller values) and can be found in the Appendix (Table A5.1).

For the Magnitude comparison task (MC), children were asked to point at the larger of two different Arabic numbers, each pair presented on a separate sheet of paper. At T1, children were presented with blocks of 6 double-digit numbers, three-digit numbers, and four-digit numbers each, resulting in 18 items. From T2 onwards, children were presented with 8 stimuli per block, resulting in 24 items. Half of the items in each block were compatible (larger unit digit within the pair belongs to larger number) and half were incompatible (larger unit digit within the pair belongs to smaller number; see Nuerk, Weger, & Willmes, 2001, for an explanation of the compatibility effect), with the overall magnitude of the compatible and the incompatible number pairs matched per block. Like for RN, overall processing time per child was measured in seconds.

Self-rating questionnaire for evaluation of mathematics
The math anxiety questionnaire used in this study was the German adaptation of the MAQ (Thomas & Dowker, 2000). In a standardization study published in German, the internal consistency (Cronbach's alpha) was reported to vary between .83 and .91 for the whole questionnaire depending on the age group examined (Krinzinger et al., 2007). The MAQ requires children to answer four different types of questions (A: 'How good are you at...?', B: 'How much do you like...?', C: 'How happy or unhappy are you if you have problems with...?', D: 'How worried are you if you have problems with...?') on one training situation (writing; e.g., 'How good are you at writing?') and on seven subsequent math related situations each (math in general, written calculations, mental calculations, easy calculations, difficult calculations, math homework, listening and understanding during math lessons; e.g., 'How much do you like math in general?'). Children were asked to mark their respective answers on a 5-point scale using different pictures for each type of

question (see Figure 5.1: check marks and crosses for very good to very bad self-perceived performance, wasps and candies for very negative to very positive attitudes, happy and unhappy faces for poor-performance unhappiness, and worried and relaxed faces for anxiety; Krinzinger et al., 2007).

The rating varied from 0 for the most negative possible answer to 4 for the most positive possible answer, thus resulting in an overall minimum score of 0 (most negative) and an overall maximum score of 28 (most positive).

For each different type of situation, children were asked to mark their respective answers to the four related questions in a different colour stated by the experimenter. The MAQ was administered in groups of 15-20 children.

Figure 5.1: MAQ pictorial rating scales for the four types of questions: A: "How good are you at...?" (very good – very bad); B: "How much do you like...?" (not at all – very much); C: "How happy or unhappy are you if you have problems with...?" (very happy – very unhappy); D: "How worried are you if you have problems with...?" (very worried – very relaxed; Krinzinger et al., 2007, copyright by courtesy of Hans Huber Verlag)

In the above mentioned standardization study (Krinzinger et al., 2007) we employed Faceted Smallest Space-Analysis (FSSA; Shye, 1985; Shye & Elizur, 1994) to examine the empirical structure underlying the four different types of questions (see 2.4.2). We found that items belonging to the first two types of questions (A: 'How good are you at...?';

B: 'How much do you like...?') as well as those tapping the second two types of questions (C: 'How happy or unhappy are you if you have problems with...?'; D: 'How worried are you if you have problems with...?') were located close together and well separated from each other, respectively. We interpreted the first aspect as general math-related attitudes (evaluation of mathematics) and the second factor as negative emotions and anxiety concerning mathematics (math anxiety).

For the present study, we will only use the first two types of questions for the assessment of children's evaluation of mathematics.

Visual-spatial working memory capacity tasks

A forward and a backward version of the Corsi-Block-Tapping task (Corsi, 1972) were administered to the children. In this task, children were presented with nine cubic blocks on a board and asked to repeat sequences of tappings on these blocks as soon as the experimenter has finished them. In both the forward and the backward version, children were presented with increasingly long sequences. Two different sequences were presented for each length, and the task was terminated after a child was not able to repeat both sequences of one length correctly.

Visual-spatial abilities tasks

To assess general visual-spatial abilities, the Beery-Buktenica Developmental Test of Visual-Motor Integration (VMI; Beery, 1997) with its supplemental Developmental Test of Visual Perception (VP; Beery, 1997) were administered. The VMI consists of 27 increasingly complex geometrical figures that children have to copy (3 on one page each). In the VP, children are presented with the same figures (but much smaller: up to 14 on one page each). Beneath these figures, up to six visually similar distractors and always one match were presented as well. Children were requested to mark the matching figures. For both tasks, correctly solved items were coded with one, incorrect items with zero.

5.2.4 Model specification

In order to examine the possible interrelations between the factors Sex, evaluation of mathematics (Eval), visual-spatial working memory capacity (VS WM), general visual-spatial abilities (Vis-Spa), and multi-digit number processing (MuDi), we applied Structural

Equation Modelling (using AMOS 7.0; see also 2.2.4). This analytical method allows not only for applying confirmatory factor analyses, but also for testing the adequacy of a set of postulated multivariate regression equations between so called observed variables/indicators (measurement model) and latent variables/factors (structural model). Latent variables can be interpreted as the commonality shared between all observed variables the factor is loading on. For example in our case, the latent variable loading on RN, WN, and MC was multi-digit number processing (MuDi). We used the maximum-likelihood method for parameter estimation (Hoyle, 1995) based on the correlation matrix of observed variables.

Measurement model

As indicators for the latent variable MuDi we used the three observed variables RN, WN, and MC. To avoid ceiling effects, we calculated the score of correctly solved problems per minute for RN and MC per time point and child and used them as indicators. As explained above, due to group testing we could not assess overall processing time for the writing Arabic numbers task at T3. Hence, the number of correctly solved problems was taken as indicator for WN. As stated above, the latent variable MuDi can be interpreted as the commonality between these three indicators – therefore the different operationalizations of observed variables should be of no concern. Mean scores for the first two questions of the MAQ (A & B; Thomas & Dowker, 2006) were used as observed variables for the latent variable Eval (see also Krinzinger, Kaufmann, & Willmes, 2009). For the latent variable VS WM the sum scores of the forward and backward versions of the Corsi-Block-Tapping task were used, and the sum scores of the VP and the VMI for the latent variable Vis-Spa.

Structural model

Our postulated regression paths between the latent variables were set according to the regular developmental sequence. This means that only paths from factors at one time point to the next and between factors at one time point were allowed, but not from factors of a later time point to an earlier one (see Figure 5.2). As we were interested in the predictors for MuDi, we allowed only for paths from all other latent variables to MuDi, but not vice versa. As sex can only influence the other latent variables and not vice versa, only respective regression paths were allowed. No regression paths were allowed between the possible mediators (Eval, VS WM, Vis-Spa) of gender differences in MuDi in the default model, as we had no specific respective hypotheses. The model specification of the structural model with all latent variables and all paths we allowed for (default model) can be seen in Figure 5.2.

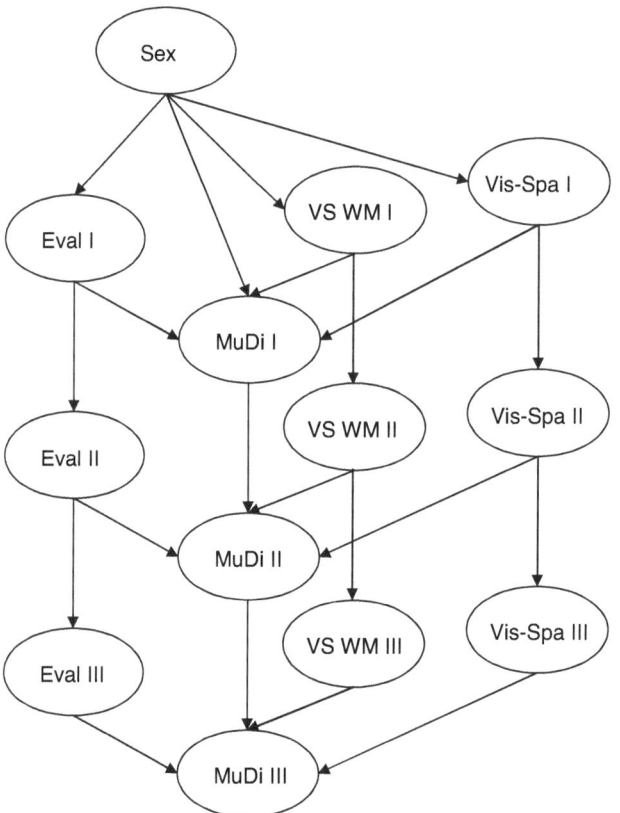

Figure 5.2: Model specification (structural part only) for a longitudinal model linking *Sex*, *evaluation of mathematics* (*Eval*; two observed variables at each time point), *visual-spatial working memory capacity* (*VS WM*; two observed variables at each time point), *general visual-spatial abilities* (*Vis-Spa*; two observed variables at each time point), and *multi-digit number processing* (*MuDi*; three observed variables at each time point) for primary school children from the end of first grade (T1) to the middle of second grade (T2)

5.2.5 Fit statistics

The same fit statistics as in Study 1 were used (see 2.2.6). Nested models with additional constraints are preferred over the unspecific default model if the difference in X^2 between the two models is either not significant or significantly better for the nested model.

5.3 Results

5.3.1 Descriptive statistics

For descriptive statistics on the different tasks at the four time points, see Table 5.1. In general, we could observe that over time, children improved their performance in all tasks, and their evaluation of mathematics decreased. This observation could be confirmed by testing for a linear trend in repeated measures ANOVAs for each indicator (all $F(1, 139) > 24.4$, all $p < .001$). The correlations between tasks can be found in the Appendix (Tables A5.2-A5.7).

5.3.2 Longitudinal model

To test for the possible specific causal relations between Sex, Eval, VS WM, Vis-Spa, and MuDi as outlined above, we set up one single longitudinal structural equation model as described in the methods section.

During the process of evaluating model fit, modification indices suggested allowing for several error covariances between observed variables. Correlated error covariances between different measures of the same construct are a common finding in longitudinal structural designs (Schumacker & Lomax, 2004). These error covariances absorb the dependencies between different measures of a certain latent construct which are elicited by the serial measurement. We allowed only for those thirteen error covariances which were suggested between observed variables loading on the same latent variable. This procedure increased model fit significantly (difference in degrees of freedom: 12, difference in X^2: 228, $p < .001$; nested model A).

Setting eight non-significant regression paths between latent variables to zero did not change model fit significantly (difference in degrees of freedom: 8, difference in X^2: 7, p = .521; nested model B).

Furthermore, three additional paths between latent variables were suggested by the modification indices, namely from Vis-Spa to VS WM and from VS WM to Eval at the first time point and from Sex to Eval at the second time point. Compared to nested model with twelve error covariances and eight non-significant regression weights set to zero, adding these three paths increased model fit again significantly (difference in degrees of freedom: 3, difference in X^2: 47, p < .001). Therefore, we selected the latter model and will present only the standardized solution for this model.

Table 5.1: Descriptive statistics of the observed variables of Study 4 (tasks written in bold font are used as observed variables in the model) at all three time points (T1: end of first grade; T2: middle of second grade; T3: end of second grade)

Task	Time point	Items	Mean	SD	Range	Skewness (SE=.205)	Kurtosis (SE=.407)
MAQ A: sum	T1	7	21.8	4.2	11-28	-0.3	-0.7
	T2	7	21.8	4.8	9-28	-0.6	-0.1
	T3	7	20.4	4.9	6-28	-0.4	0.0
MAQ B: sum	T1	7	20.5	5.6	5-28	-0.6	-0.3
	T2	7	20.0	5.9	5-28	-0.4	-0.6
	T3	7	17.9	6.1	0-28	-0.3	-0.4
Corsi-Block forward: sum	T1	16	9.8	2.0	5-14	0.0	-0.6
	T2	16	10.6	2.2	5-15	-0.1	-0.3
	T3	16	10.4	2.1	6-15	0.1	-0.5
Corsi-Block backward: sum	T1	16	8.6	2.1	4-14	0.4	0.1
	T2	16	9.4	2.1	4-15	0.1	-0.2
	T3	16	9.8	2.0	5-14	0.0	-0.1
VMI: sum (copying figures)	T1	27	19.9	2.7	9-27	-0.1	0.7
	T2	27	20.9	2.2	14-27	-0.1	0.6
	T3	27	20.9	2.4	14-27	-0.3	0.2
VP: sum (matching figures)	T1	27	22.1	2.3	14-27	-0.5	0.3
	T2	27	22.3	2.1	17-27	-0.0	0.2
	T3	27	22.9	2.2	17-27	0.0	2.3
Writing Arabic	T1	30	18.6	8.3	5-30	0.1	-1.5

numbers:	T2	38	26.6	11.3	2-38	-0.7	-0.8
sum	T3	38	30.8	9.4	5-38	-1.4	0.7
Reading Arabic	T1	30	17.1	7.7	7-30	0.5	-1.4
numbers: sum	T2	38	29.7	9.5	3-38	-1.0	-0.1
	T3	38	32.5	7.4	6-38	-1.7	2.4
Reading Arabic	T1	30	163.6	84.5	20-420	0.7	0.0
numbers: time	T2	38	364.7	96.1	114-631	0.3	0.3
in sec	T3	38	154.1	80.4	56-597	2.2	7.0
Reading	T1	30	9.4	8.2	2-35	1.3	0.5
Arabic	T2	38	12.6	7.8	1-34	0.6	-0.2
numbers: corr	T3	38	16.4	8.4	1-41	0.3	-0.3
per min							
Magnitude	T1	18	14.5	3.6	7-18	-0.5	-1.1
comparison:	T2	24	22.7	2.1	14-24	-2.4	5.5
sum	T3	24	22.9	2.0	14-24	-2.6	6.9
Magnitude	T1	18	53.9	25.6	27-214	3.1	13.8
comparison:	T2	24	47.2	12.0	26-104	2.0	6.1
time in sec	T3	24	42.9	10.7	29-106	2.8	11.7
Magnitude	T1	18	19.0	8.6	3-40	0.3	-0.5
comparison:	T2	24	30.5	7.3	10-53	-0.4	1.2
corr per min	T3	24	33.6	7.2	10-48	-0.8	0.8

The fit statistics for the default model, the two nested models and the selected model can be obtained from Table 5.2. Four out of five different fit statistics can be considered satisfactory for the selected model, indicating that our model fits the data well.

The standardized structural solution for the selected model can be seen in Figure 5.2.

Overall, it can be seen that the regression weights from one latent variable to the same latent variable half a year later are very high (> .70 for Eval, .80 - .99 for the cognitive factors), indicating stable development. Significant predictors for MuDi at the first time point were Eval, Vis-Spa, and Sex.

The relation between the possible mediators of gender differences of MuDi can be characterized as follows: Vis-Spa was a predictor for VS WM, which in turn predicted Eval. Sex influenced MuDi at the first time point and Eval at the first and the second time point. VS WM was not a significant predictor for MuDi at any time point. Neither Vis-Spa nor VS WM were predicted by Sex at any time point. At the second time point (middle of second grade), the only additional variance in MuDi (which was not already explained at the first time point) was explained by Eval. At the third time point, no additional variance of MuDi was explained by any other latent variable. The factor loadings (standardized regression weights) of the latent constructs on the observed variables for the selected model can be obtained from Table 5.3.

Table 5.2: Fit statistics of the different models: Unspecific default model, nested model A, nested model B, and the selected nested model

Fit statistic	Default model	Nested model A	Nested model B	Selected nested model
X^2	663	435	443	396
Degrees of freedom (df)	330	318	326	323
X^2 / df	2.0	1.4 [a]	1.4 [a]	1.2 [a]
AGFI	.693	.782	.785	.804
CFI	.852	.948	.948	.968 [a]
RMSEA	.085	.051	.051	.040 [a]
AIC	815	611	602	562 [a]

[a] This value is conventionally considered to indicate satisfactory model fit.

Table 5.3: Standardized regression weights of observed variables of Study 4

Latent variable	Observed variable	Standardized regression weight at T1	Standardized regression weight at T2	Standardized regression weight at T3
Eval	MAQ A: sum	.93	.98	.94
	MAQ B: sum	.74	.80	.69
VS WM	Corsi-Block forward: sum	.60	.69	.77
	Corsi-Block backward: sum	.71	.67	.68
Vis-Spa	VMI: sum	.71	.72	.62
	VP: sum	.58	.69	.57
MuDi	WN	.91	.78	.71
	RN	.99	.95	.94
	MC	.63	.55	.61

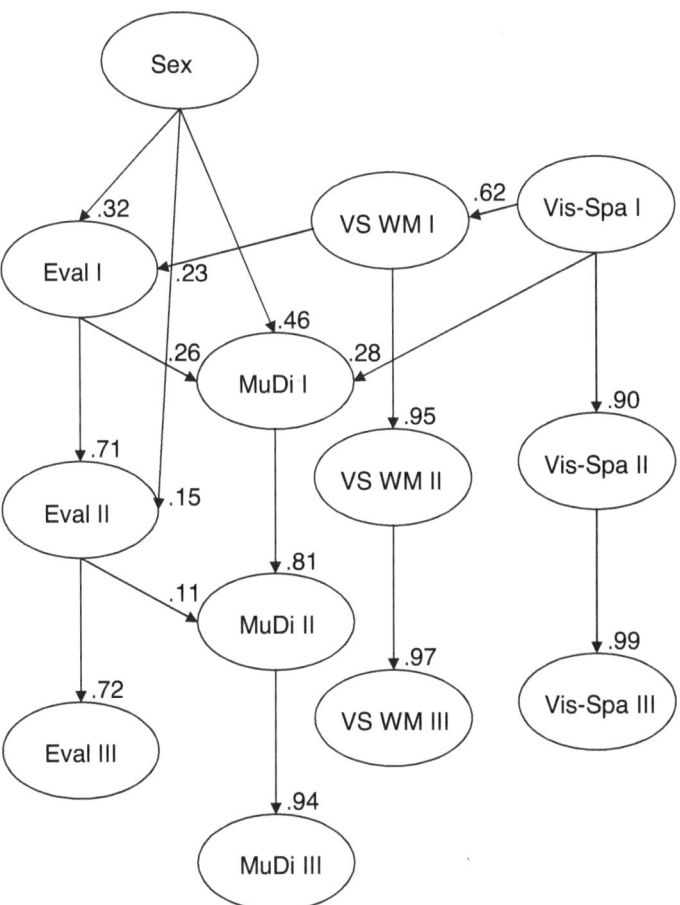

Figure 5.3: Selected nested longitudinal model (standardized solution, structural part only) for linking *Sex*, *evaluation of mathematics* (*Eval*; two observed variables at each time point), *visual-spatial working memory capacity* (*VS WM*; two observed variables at each time point), *general visual-spatial abilities* (*Vis-Spa*; two observed variables at each time point), and *multi-digit number processing* (*MuDi*; three observed variables at each time point) for primary school children from the end of first grade (T1) to the middle of second grade (T2)

5.4 Discussion

In this study, we could replicate the gender differences in multi-digit number processing. These gender differences were partly mediated by children's evaluation of mathematics at the end of first grade and at the middle of second grade, but this mediation was not present any more at the end of second grade. Furthermore, at the first time point (end of first grade) the effect of evaluation of mathematics on multi-digit number processing was mediated by visual-spatial working memory capacity, which was mediated by general visual-spatial ability. Yet, we found no direct influence of visual-spatial working memory on multi-digit number processing. What is more, we found no gender differences in general visual-spatial abilities. Yet, this factor was still a significant predictor for multi-digit number processing, which means that both within boys and within girls children with better visual-spatial abilities were also better in multi-digit number processing.

Furthermore, we found stable development of all latent variables tested in our sample from second grade onwards, which may be seen as an indicator for the importance of early intervention.

In summary, we found that gender had the highest impact on multi-digit number processing, followed (and partly mediated) by evaluation of mathematics as well as by visual-spatial abilities, whereas visual-spatial working memory capacity was no significant predictor. These results will be discussed in more detail.

5.4.1 Why are visual-spatial abilities important for the acquisition of multi-digit number processing?

In general, it is important to explain how better visual-spatial abilities could help children in learning how to read, write, and compare multi-digit Arabic numbers. Most studies showing mediation of gender differences in mathematics by visual-spatial abilities were concerned with general mathematical problem solving abilities or mathematics standard achievement scores (Casey et al., 1995, 2001; Fennema et al., 1998; Rosselli et al., 2008; Gallagher et al., 2000), not multi-digit number processing. For the latter, children have to acquire a basic understanding of the regularity of the Arabic notational system of numbers for successful multi-digit number processing (Barrouillet et al., 2006). As this regularity has inherent spatial features (as the term "place-value system" indicates), it is plausible that a preference for spatial thinking strategies or better visual-spatial abilities should provide individuals with an advantage for acquiring the place-value system of Arabic numbers. For example, Geary (1993) describes a specific subtype of DD where children have mainly deficits in the visual-spatial representation of numerical information. He compares this subtype of DD with spatial acalculia, which is often associated with right posterior lesions in neuropsychological patients (e.g., Hartje, 1987), and lists "the misalignment of numbers in multicolumn arithmetic problems, [...] and difficulty with place value and decimals" (Geary, 1993, p. 352) as typical problems of both syndromes.[3]

5.4.2 Why is visual-spatial working memory capacity not a predictor for multi-digit number processing?

Contrary to general visual-spatial abilities, the capacity of visual-spatial working memory was no significant predictor for multi-digit number processing abilities in our sample, even though a recent study by Zuber and colleagues (2009) suggested so. As the added regression path in our selected model from visual-spatial abilities to visual-spatial working memory capacity showed, the finding by Zuber and colleagues (2009) may well be explained as visual-spatial working memory being a mediator of general visual-spatial abilities. This could mean that the better the visual-spatial abilities, the higher the respective working memory capacity – which has in itself no direct impact on multi-digit number processing. Of course, this finding has to be replicated in future studies. Furthermore, we did not include measures for the central executive aspect of working memory (Baddeley, 2002; Baddeley & Hitch, 1974) and can therefore make no statement on its importance for multi-digit number processing.

---

[3] Concerning the observed gender differences in Multiplication in Study 3a, spatial problem solving strategies might also play a supporting role when children have very little or no formal experience with this new type of arithmetic operation and have not yet learn the multiplication tables by heart, which is the case for our sample. It may be speculated that this initial male advantage in solving multiplications could even be the cause for the better fact retrieval abilities that have been reported for male adult subjects (Royer et al., 1999). Finally, because even the multiplication task is influenced by place-value information (consistency effect: see Domahs, Delazer, & Nuerk, 2006; Domahs, Domahs, Schlesewsky, Ratinckx, Verguts, Willmes, & Nuerk, 2007; Verguts & Fias, 2005), better place-value understanding may also aid performance in the multiplication tasks.

5.5 Conclusion

To summarize, the interesting gender differences in our studies could potentially be explained by a better place-value understanding in boys. This male advantage may at least partly be based on better visual-spatial abilities in boys, even though we found no gender differences in our tasks tapping visual-spatial ability in this study. As the gender differences in our study could not be fully explained (as the significant direct influence of sex on multi-digit number processing shows), we propose that it is a worthwhile endeavour for future studies to examine which other specific spatial abilities (e.g., three-dimensional processing) or which specific socio-cultural factors (e.g., teaching styles) that are not as general as the Gender Gap Index (Hausmann et al., 2006) may influence what kinds of specific numerical representations (e.g., place-value understanding).

In general, we think that our results of complex mediating effects between cognitive abilities and affective states like evaluation of mathematics on gender differences in multi-digit number processing provide a good example for the psychobiosocial model of gender differences in mathematics (Halpern, 1997). This is to say that neither socio-cultural factors alone nor biological factors alone will be sufficient to fully explain gender differences in numerical abilities.

6 Study 5: Multi-digit number processing as predictor of multi-digit calculation performance and an exact number magnitude representation

6.1 Introduction

Within the domain of numerical cognition, the acquisition of multi-digit number processing seems to be one of the most challenging numerical abilities children have to learn (Geary, 2000). Children need several years of training to master the meaning of the decimal Arabic number system and the rules for transcoding multi-digit numbers from one format to another (e.g., verbal to Arabic or vice versa; Noël & Turconi, 1999; Power & Dal Martello, 1990; 1997; Seron & Fayol, 1994), especially in intransparent number word systems such as German (Zuber et al., 2009; see also Study 2).

6.1.1 Models of number processing

In general, adult transcoding models can be grouped in two main categories. They either propose that transcoding numerical information from one format to another (e.g., from spoken number words to Arabic numerals) is a semantic (for a review about semantic models see McCloskey, 1992) or an asemantic (e.g., Deloche & Seron, 1987) process.
The only developmental transcoding model that has been published hitherto belongs to the second group. This developmental, asemantic, and procedural model for transcoding from verbal to Arabic numerals (ADAPT; Barrouillet et al., 2004) proposes that the transcoding process happens in purely algorithmic form and that it does not require a semantic representation of the quantity represented in the source code. Furthermore, it states that the simplest and most frequent numerals with only few lexical units, but not three-digit and higher numbers (such as 378 or 4185), will soon be lexicalized. This means that for double-digit numbers, the transcoding process very early shifts from algorithmic strategies to the direct retrieval of digital forms from long-term memory. Interestingly, even though the ADAPT model explicitly states that their proposed procedural system is acquired through learning and practice and does not require semantic understanding, the authors acknowledge in the last sentence of their manuscript that "the meaningful structure of the

digital forms produced by this system would provide children with a basis for understanding how numbers are constructed, what they represent, and how they can be used" (Barrouillet et al., 2004, p. 392).

This statement already suggests that understanding the base-10 system of multi-digit Arabic numbers may play an important role for different aspects of numerical cognition. For example, in primary school children of different ages, the task Transcoding of number words to Arabic numerals as described above loaded on the same component as Reading of Arabic numerals, Magnitude comparison of number words, and Magnitude comparison of Arabic numerals, whereas the calculation tasks Addition, Subtraction, Multiplication, Word problems, and Conceptual understanding loaded on another component (see 2.3.3). A purely syntactical and rule-based transcoding process of number words can not explain why performance on different tasks using multiple multi-digit number modalities are all explained by one single cognitive component, but the notion of a base-10 concept can. Concerning the development of respective base-10 concepts, two theories have been proposed.

The more general theory by Case and his collaborators emphasizes that central conceptual structures represent children's core knowledge in a domain, and that they can be applied to a full range of different tasks (Case, 1996). These conceptual structures are thought to become increasingly complex during development (from unidimensional to bidimensional to integrated bidimensional thought) by a reciprocal process of general conceptual insights and more specific task understanding. In the number domain, these stages are bringing together analogue magnitude understanding and counting knowledge into a mental number/counting line (unidimensional), using two mental number lines (e.g., for ones and tens; bidimensional), and understanding relations between different mental number lines (integrated bidimensional; Okamoto & Case, 1996).

Similarly, Fuson and her colleagues differentiate between different conceptual structures of double-digit numbers that evolve during development, namely first a unitary counting-string based representation, second three different possible non-integrated two-dimensional concepts, and third the integrated two-dimensional concept (Fuson et al., 1997).

Both models predict that the conceptual structures children have developed for double-digit numbers will influence their abilities in other numerical tasks as well. Recent studies on the development of numerical competencies during early primary school years foster this view. For example, magnitude comparison of double-digit numbers seems to be influenced by an increasing integration of unit- and decade digits with age (Nuerk et al., 2004; Pixner et al., 2009; but see Ashkenazi, Mark-Zigdon, & Henik, 2009, for different findings).

Specifically, multi-digit calculation (Fuson et al., 1997; Miller et al., 1995) and the development of an exact, linear magnitude representation for large numbers (Booth & Siegler, 2006; Dehaene, 2007; Laski & Siegler, 2007; Opfer & Siegler, 2007; Siegler & Booth, 2004; Siegler & Opfer, 2003; von Aster & Shalev, 2007) are thought to depend on understanding the base-10 system of multi-digit numbers (Moeller, Pixner, Kaufmann, & Nuerk, 2009).

In the following, we will first review the state of research on these topics. Then, we will present a longitudinal structural equation model showing that multi-digit number processing which is thought to rely on base-10 understanding is a significant predictor of a linear number magnitude representation and multi-digit calculation skills.

6.1.2 The role of multi-digit number processing in the development of an exact, linear number magnitude representation

Recent hypotheses state that during development, the acquisition of numeric symbols will change the form of the spatial left-to-right oriented magnitude representation of numbers (so called "mental number line"; Dehaene, Bossini, & Giraux, 1993) from a logarithmic function for approximate, non-symbolic numerosities to an exact, linear function for symbolic numbers.

This hypothesis is supported by the empirical findings of Siegler and his co-workers. Siegler and colleagues also predict a developmental change from an initial and intuitive logarithmic number magnitude representation to a more advanced and exact linear representation, which is superior in supporting accurate estimation and calculation for larger numbers. Using a number line estimation task (e.g., presenting an empty physical horizontal line with the labelled endpoints "0" and "100" and asking children to indicate with a hatch-mark where a specific number like "27" goes in between the two endpoints on the line) in different cross-sectional samples of young children, they report an increasing linearity of number representations for the number range 0-100 between kindergarten and second grade (Booth & Siegler, 2006; Laski & Siegler, 2007; Siegler & Booth, 2004) and between second and fourth grade for the number range 0-1000 (Opfer & Siegler, 2007;

Siegler & Opfer, 2003; but see Ebersbach, Luwel, Frick, Onghena, & Verschaffel, 2008; Moeller et al., 2009, for different non-logarithmic views about the initial representational functions).

Parallel developmental trends towards exact estimation patterns and consistent individual differences in three different numerical estimation tasks in the same number range (number line estimation, numerosity estimation, and computational estimation; Booth & Siegler, 2006) and even in different numerical tasks (numerical categorization, number line estimation, and numerical magnitude comparison; Laski & Siegler, 2007) support the notion that the number line estimation task actually figures children's number magnitude representations and can not only be considered a spatial task.

The above cited studies explain the observed shifts to more accurate linear functions in children's estimation patterns by their increasing experience with larger numbers (e.g., Siegler & Booth, 2004; Ebersbach et al., 2008). Yet, this experience or familiarity with larger number ranges is usually not well specified. The most likely meaning of familiarity with a specific number range may be the ability to read and write the respective numbers fluently, i.e. number symbol knowledge. In a recent paper on number representation, Dehaene (2007) explicitly argues for a representational change in numerical magnitude representation from a logarithmic to a linear function via the acquisition of symbolic numbers. He bases his assumptions on the on the neural network model by Verguts and Fias (2004), which encodes non-symbolic numerosities logarithmically but symbolic numerosities linearly and with fixed variability. If this assumption is correct, fluency in reading and writing numbers in a specific range should be sufficient for an exact, linear representation of these numbers.

A recent developmental model of number representation by von Aster and Shalev (2007) takes one theoretical step forward and predicts that the acquisition of the Arabic number system as such is a prerequisite for the development of a linear "mental number line". In line with this hypothesis, Laski and Siegler (2007) explained the development towards a linear number magnitude representation with "increasing conceptual understanding of the number system" (p. 1726). Accordingly, Moeller and colleagues (2009) interpret their findings of increasing linearity in number representations as being caused by a better integration of units and decades into the Arabic base-10 system with development. This would mean that general understanding of the base-10 system, which is needed for general multi-digit number processing, should be a predictor for an exact number representation. Yet, systematic longitudinal empirical evidence for this assumption has to our knowledge not been published.

6.1.3 The role of multi-digit number processing in learning multi-digit calculation

Another numerical competency children have to acquire during primary school is multi-digit calculation. In Germany, children gain first experience with double-digit additions and subtractions during second grade.
According to Baroody (2003), the flexible application of arithmetic knowledge (e.g. for calculation) depends on the integration of procedural knowledge ("knowing how to": application of carry- and borrow-procedures) and conceptual knowledge ("knowing why": understanding arithmetic principles). Rittle-Johnson and Siegler (1998) even state that understanding arithmetic procedures requires understanding of the concept of place-value. In a longitudinal study about the developmental relationship between procedural and conceptual knowledge, Hiebert and Wearne (1996) found that children who had good conceptual understanding of base-10 in first grade steadily increased their procedural skill in arithmetic over time, whereas non-understanders had much flatter learning curves. Furthermore, most of the highly procedurally skilled children in fourth grade had been classified as understanders in first grade. Resnick (1992) admitted that the successful application of an arithmetic procedure is possible without understanding the concepts behind it (base-10 system and concept of addition or subtraction), but she also concluded that difficulties in learning are mostly a result of incomplete connections between syntactic rules and semantic knowledge, leading to systematic errors in performance ("buggy algorithms"). Furthermore, Fuson and colleagues (1997) show in their developmental model about conceptual structures of two-digit numbers that the respective concepts children have directly influence which methods of multi-digit addition and subtraction they are able to use. Therefore, it is to be expected that multi-digit number processing will be an important predictor for double-digit calculation performance.

6.1.4 Developmental influences between multi-digit calculation performance and linearity of number magnitude representation

As the development of a linear number representation in the range 0-100 occurs at about the same time as children gain first experience with addition and subtraction in the respective number range (until the end of second grade), it is reasonable to expect that these two numerical competencies may not develop independently. Interestingly, the question of possible influences between the linearity of number representation and calculation performance has been discussed from both perspectives.

Chronologically, the first hypothesis was brought up by Petitto (1990), who stated that experience with calculation in a specific number range would lead to the discovery of the proportional relations between numbers and to the implementation of these relations into (then exact) mental representations, which had covered only the sequential relations of numbers before.

Even more informative is a recent study by Dehaene, Izard, Spelke, and Pica (2008) who examined the numerical representations of Amazonian Indigene children and adults. They found that only individuals with at least three years of mathematical education showed a linear representation for Portuguese number words. The authors speculate that calculation experience may be crucial for developing a linear number representation.

As already mentioned before, the reverse influence of a linear number representation facilitating exact calculation with larger numbers via better estimation abilities is supported by Booth and Siegler, who frequently found positive correlations between the linearity of number representation and math achievement test scores in two different studies (Booth & Siegler, 2006; Siegler & Booth, 2004).

Yet, none of the above mentioned studies concerning the possible impact of accurate number representations on calculation skills or vice versa has examined the impact of base-10 understanding as a possible predictor for both numerical skills.

6.2   Objectives of Study 5

The first research question we wanted to address is whether number symbol knowledge in a specific range is sufficient for the development of a linear and therefore exact number magnitude representation (Dehaene, 2007; Verguts & Fias, 2004). If this assumption is true, flawless transcoding of double-digit numbers should automatically lead to a linear representation for this range.

Furthermore, we wanted to examine the developmental influences between multi-digit number processing, linearity of number magnitude representation in the range 0-100, and calculation performance in the same number range using a longitudinal design. Specifically, we aimed at finding out whether in the number range 0-100

(i) processing of three- and four-digit numbers helps in acquiring an exact, linear number representation,

(ii) processing of three- and four-digit numbers is predictive for acquiring double-digit calculation skills, and whether

(iii) an exact number representation influences calculation performance or vice versa.

6.3 Methods

6.3.1 Participants and general procedure

The same sample as for Study 4 (see 5.2.1) was tested. For this study, we used the same tasks as in Study 4 for multi-digit number processing (see 5.2.3), but additionally a Number line estimation task and four Double-digit calculation tasks (from T2 onwards) were used as well. All new tasks were administered at a fourth time point as well (T1: end of first grade; T2: middle of second grade; T3: end of second grade; T4: middle of third grade). The Number line estimation task was always presented in individual testing sessions, whereas the Double-digit calculation tasks were assessed in groups of 15-20 children.

6.3.2 Stimuli and task procedures

Reading, writing and comparing Arabic numbers tasks
The same stimuli and task procedures as for Study 4 (see 5.2.3) were used.

According to the ADAPT model (Barrouillet et al., 2004), we will assign different processes to the transcoding of double-digit vs. three- and four-digit numbers. On the one hand, familiarity with double-digit numbers will lead to their lexicalized retrieval from long-term memory. On the other hand, learning how to transcode three- and four-digit numbers will trigger understanding the "meaningful structure" underlying them. In line with this differentiation, we will define fluent processing of double-digit numbers as number symbol knowledge in this range, whereas the commonality in processing of three- and four-digit numbers in all three tasks will be the operationalization of more general base-10 understanding of multi-digit numbers.

Number line estimation task

In the Number line estimation task (a variant of the well-established task by Siegler & Opfer, 2003), children were shown a 10 cm long physical line with the labelled end-points 0 and 100 and asked: "Here you see a number line which goes from 0 to 100. If 0 is here [point to left end] and 100 is here [point to right end], where does 50 lie between 0 and 100? Please mark the position of 50 with a hatch-mark!" After this instructional trial, children were presented with 18 more 10 cm long lines one at a time and were asked the same question for all decade numbers (only from T2 onwards) and the same number of non-decade numbers (for all four time points) in pseudo-randomized order.

Double-digit calculation tasks

Double-digit calculation performance was measured with four different tasks: Additions without carrying, Additions requiring a carry-procedure, Subtractions without borrowing, and Subtractions requiring a borrow-procedure. For each task, one block of 10 different problems was presented to the children. The stimuli were only double-digit numbers (without decade numbers), and the results of the calculations lay between 1 and 100. As children were taught double-digit calculations only just before T2 (middle of second grade), we did not administer them at T1. Children were always given a specific time interval to work on one calculation block: three minutes at T2, one minute at T3, and half a minute at T4. Due to the small number of items per block and the timed task procedure, we did not match the stimuli across blocks for problem size.

6.3.3 Model specification

To examine the possible interrelations between the three factors multi-digit number processing, linearity of number representation, and double-digit calculation, we applied again Structural Equation Modelling (using AMOS 7.0; see also 2.2.4). We used the maximum-likelihood method (Hoyle, 1995) based on the correlations between observed variables which can be obtained in the Appendix (Tables A5.7 and A6.1-6.5).

Measurement model
As indicators for the latent variable multi-digit number processing (MuDi) we used the same three observed variables as for Study 4 (see 5.2.4), namely Reading, Writing, and Comparing Arabic numbers.

The latent variable linearity of number representation (Lin) was assessed using a number line estimation task for the range 0-100. For each child, the slopes of the linear functions for the given estimates in mm for decade numbers and non-decade numbers were calculated separately. Unfortunately, we presented children only with the non-decade numbers at T1. Therefore we calculated two slopes for half of these trials each (numbers 2, 27, 47, 64, 86 and 13, 35, 52, 75, 99 respectively) to end up with two observed variables for Lin at T1 as well. A slope of 1 indicates exact linearity in the number line task because it indicates that e.g. the number 20 is marked where the number 20 should be. Deviations from a slope of 1 imply inadequate linearity. Therefore, the absolute deviations from 1 of the individual slopes were calculated and taken as indicators for the latent variable linearity of number representation instead of the slopes themselves[4].

The numbers of correctly solved problems per minute and block per examination were chosen as the four different observed variables loading on the latent variable double-digit calculation performance (DD C).

Structural model

---

[4] We decided not to use the respective explained variance of the linear slope (R^2) for three reasons. First, a purely technical reason is that the R^2's show strong ceiling effects and deviate substantially from a normal distribution (see Results section and Table 3). Second, a theoretical reason is that the slope is a more direct measure of the property to be indexed than R^2. Third, high correlations between the two respective measures for all eight indicators but one (all $r > .72$, all $p < .001$ - despite the ceiling effects of the R^2's) support the validity of the slope as indicator for the linearity of number representation.

Our postulated regression paths between the latent variables were set according to the regular developmental sequence. This means that only paths from factors at one time point to the next and between factors of one time point were allowed, but not from factors of a later time point to an earlier one (see Figure 6.1). Furthermore, MuDi has only been mentioned as predictor of and not as being influenced by Lin or DD C in the literature. Therefore, we only set paths pointing from MuDi to the other two latent variables and not vice versa. Regarding the possible influence between Lin and DD C, we allowed paths from each latent variable pointing to the other half a year later. The causal relation between these two factors at the same time point was specified as Lin predicting DD C. The model specification of the structural model with all latent variables and all paths we allowed for (unspecific default model) can be seen in Figure 6.1.

6.3.4 Fit statistics

The same fit statistics were used as in Study 1 and Study 4 (see 2.2.6).

6.4 Results

6.4.1 Descriptive Statistics

For descriptive statistics on the different tasks at the four time points, see Table 5.1 (Writing, Reading, and Comparing Arabic numbers) and Table 6.1 (Number line estimation task and Double-digit calculation tasks).

In general, we could observe that over time, children improved their performance in all tasks. This observation could be confirmed by testing for a linear trend in repeated measures ANOVAs for each indicator (all $F(1, 139) > 41.8$, all $p < .001$). As WN showed strong ceiling effects at T4, we refrained from including the latent variable of MuDi for this time point in our model.

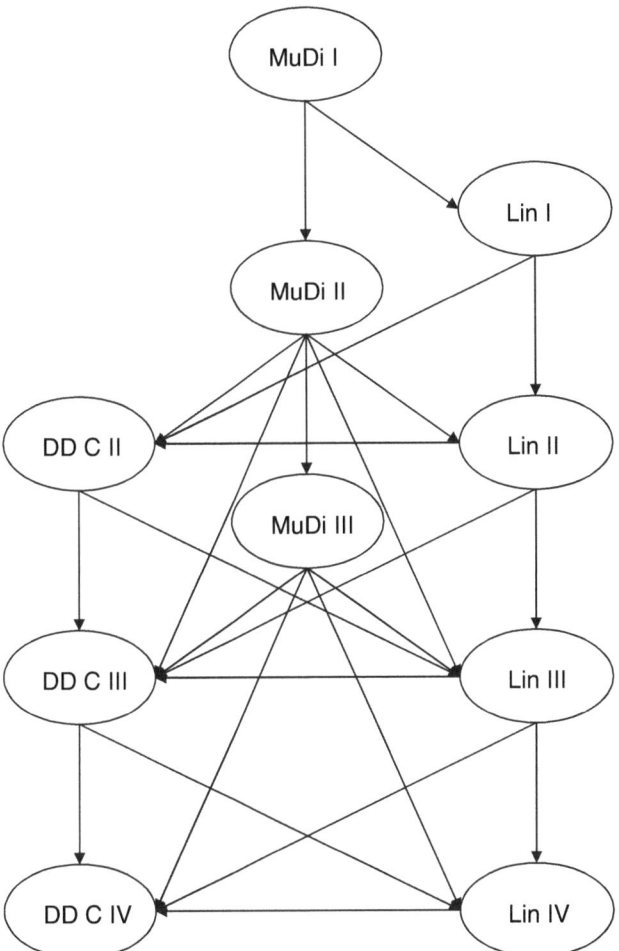

Figure 6.1: Model specification (structural part only; unspecific default model) for a longitudinal developmental model linking *multi-digit number processing* (*MuDi*; 3 observed variables each), *double-digit calculation performance* (*DD C*; 4 observed variables each), and *linearity of number representation* (*Lin*; 2 observed variables each) for primary school children from the end of grade 1 (T1) to the middle of grade 3 (T4)

To show the development of number line estimation descriptively, we plotted the mean estimates for all 19 numbers between 2 and 99 in mm (only non-decade numbers at T1) on the y-axis against the actual values of these numbers in mm on the x-axis for all four time points (see Figure 6.2). In general, the estimates are approaching an exact, linear function with development in line with the results by Siegler and Booth (2004) and Booth and Siegler (2006). The observation of increasing linearity of estimation patterns was supported by linear regression analyses using the linear function and the logarithmic function as predictors for the mean estimates for each time point. From T1 to T4, the fit of the linear function for the mean estimates of all items increased (adjusted R^2s, respectively: T1: 0.904; T2: 0.961; T3: 0.964; T4: 0.974), whereas the fit for the logarithmic function decreased with development (adjusted R^2s: T1: 0.883; T2: 0.768; T3: 0.753; T4: 0.736). The linear trend could be confirmed by applying repeated measures ANOVAs for the fit of both functions (both $F(1, 139) > 64.9$, both $p < .001$).

Furthermore, the individual number of children whose estimates were better fit by the linear function in a stepwise multiple linear regression analysis increases monotonically with development, whereas at the same time the number of children whose estimates are better fit by the logarithmic function decreases, as can be seen in Table 6.2.

Table 6.1: Descriptive statistics of the Number line estimation task and Double-digit addition (DD A) and Subtraction (DD S) tasks at all possible time points (T1: end of first grade; T2: middle of second grade; T3: end of second grade; T4: middle of third grade; tasks written in bold font are used as observed variables in the model)

Task	Time point	Items	Mean	SD	Range	Skewness (SE=.205)	Kurtosis (SE=.407)
R^2 for linear function of non-decade numbers	T1a	5	0.75	0.2	0.00-0.99	-1.1	0.7
	T1b	4	0.77	0.2	0.00-1.00	-1.6	2.2
	T2	9	0.83	0.1	0.19-0.99	-1.9	4.0
	T3	9	0.84	0.2	0.20-0.99	-2.1	4.3
	T4	9	0.88	0.1	0.31-0.99	-2.4	7.8
Slope for linear function of non-decade numbers	T1a	5	0.53	0.2	0.00-1.17	0.1	0.2
	T1b	4	0.58	0.3	-0.40-1.15	-0.8	1.5
	T2	9	0.84	0.2	0.30-1.16	-0.6	0.5
	T3	9	0.86	0.2	0.35-1.19	-0.6	0.1
	T4	9	0.90	0.1	0.33-1.17	-0.9	1.8
Absolute deviation of	T1a	5	0.48	0.2	0.03-1.00	0.2	-0.6

slope for linear function of non-decade numbers from 1	T1b	4	0.43	0.3	0.00-1.40	1.0	1.7
	T2	9	0.18	0.1	0.00-0.70	1.1	1.5
	T3	9	0.17	0.1	0.00-0.65	1.2	1.0
	T4	9	0.13	0.1	0.00-0.67	1.7	4.4
R^2 for linear function of decade numbers	T2	9	0.80	0.2	0.00-1.00	-2.0	3.3
	T3	9	0.81	0.2	0.00-0.99	-2.0	4.0
	T4	9	0.90	0.1	0.03-0.99	-3.8	18.2
R^2 for linear function of decade numbers	T2	9	0.80	0.2	0.00-1.00	-2.0	3.3
	T3	9	0.81	0.2	0.00-0.99	-2.0	4.0
	T4	9	0.90	0.1	0.03-0.99	-3.8	18.2
Slope for linear function of decade numbers	T2	9	0.76	0.3	-0.03-1.22	-0.6	0.4
	T3	9	0.79	0.3	0.02-1.25	-0.5	0.1
	T4	9	0.88	0.2	0.06-1.31	-0.8	1.5
Abs. dev. of slope for lin. function of decade numbers from 1	T2	9	0.27	0.2	0.00-1.03	1.2	1.2
	T3	9	0.26	0.2	0.01-0.98	1.2	1.6
	T4	9	0.28	0.2	0.00-0.94	1.8	4.4
DD A without carrying: corr per min	T2	[a]	2.8	0.8	0-3	-1.9	3.0
	T3	[b]	5.2	1.9	0-7	-0.9	-0.2
	T4	[c]	8.8	3.2	0-14	-0.2	-0.5
DD S without borrowing: corr per min	T2	[a]	2.2	1.0	0-3	-0.7	-0.5
	T3	[b]	3.1	1.8	0-6	0.0	-1.0
	T4	[c]	4.8	2.2	0-8	-0.3	-0.6
DD A with carrying: corr per min	T2	[a]	2.0	1.2	0-3	-0.4	-1.4
	T3	[b]	4.0	2.3	0-7	-0.3	-1.1
	T4	[c]	6.8	3.3	0-12	-0.1	-0.9
DD S with borrowing: corr per min	T2	[a]	1.1	1.1	0-3	0.6	-1.0
	T3	[b]	2.6	1.9	0-6	0.3	-1.0
	T4	[c]	4.3	2.8	0-10	0.0	-0.7

[a] 3 minutes were allowed to work on 10 items. [b] 1 minute was allowed to work on 10 items. [c] 30 seconds were allowed to work on 10 items.

Table 6.2: Absolute number and percent of individual estimation patterns fit by the linear function, the logarithmic function, both of them (mixed), or none at the different time points (T1: end of first grade; T2: middle of second grade; T3: end of second grade; T4: middle of third grade)

Number of estimates fit by function	T1	T2	T3	T4
linear (%)	73 (52%)	111 (79%)	120 (86%)	128 (91%)
mixed (%)	5 (3%)	6 (4%)	2 (1%)	1 (1%)
logarithmic (%)	52 (37%)	16 (11%)	14 (10%)	5 (4%)
none (%)	10 (7%)	0 (0%)	0 (0%)	0 (0%)
missing (%)	0 (0%)	7 (5%)	4 (3%)	6 (4%)

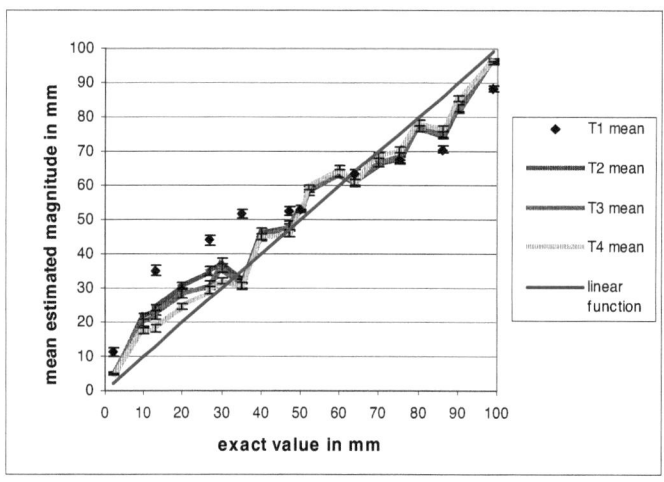

Figure 6.2: Development of the number line estimation task from T1 (diamonds: end of first grade; non-decade numbers only) to T2 (dark grey line: middle of second grade), T3 (grey line: end of second grade), and T4 (light grey line: middle of third grade) with mean estimated magnitude in mm plotted on the y-axis and the exact value in mm plotted on the x-axis. Error bars indicate standard errors; the black line shows the exact linear function

6.4.2 The role of number symbol knowledge in the number range 0-100 for the linearity of number representation in the same range

The first question we wanted to address was whether in the number range 0-100, number symbol knowledge per se is sufficient for an exact, linear number representation (Dehaene, 2007; Verguts & Fias, 2004). To test this hypothesis, we analysed the children's performance in transcoding problems for numbers smaller than 100 separately at T1 (end of first grade). At this time point, children were presented with six items in this range for reading Arabic numbers and writing Arabic numbers each. The mean percent of correctly solved items were 94% for each task (with standard deviations of 10% and 13%, respectively). Furthermore, the relative number of children who correctly solved all items smaller than 100 was 76% for writing and 72% for reading numbers. These results show that children are already very proficient in number symbol knowledge for double-digit numbers at the end of first grade. At the same time point, in about half of the children (44%) the linear function accounted for no variance at all of their number line estimation patterns (see Table 2). Furthermore, we separately analysed the fits of these children who were 100% correct in reading and writing double-digit numbers, hence showing mastery of double-digit numbers. Of the 82 children who met this criterion, in 34 (42% of this group; 24% of the whole sample) the linear function explained again no variance of their number line estimation pattern. Thus, number symbol knowledge alone is clearly not sufficient for a linear number representation in the respective number range.

6.4.3 Longitudinal model

To test for the possible specific causal relations between MuDi, Lin, and DD C as outlined above, we set up one single longitudinal structural equation model as described in the methods section. The factor loadings (standardized regression weights) of the latent constructs on the observed variables can be obtained from Table 6.3.
Correlations between error covariances as suggested by the modification indices in the process of evaluation of model fit were only allowed between observed variables loading on the same latent variable at one specific time point or between observed variables assessed by the same task on different time points. Correlated error covariances between different measures of the same construct are a common finding in longitudinal structural

designs (Schumacker & Lomax, 2004). These error covariances absorb the dependencies between different measures of a certain latent construct which are elicited by the serial measurement. Between all 29 observed variables, 16 error covariances were allowed for. The fit statistics of our model can be found in Table 6.4.

Table 6.3: Factor loadings (FL: standardized regression weights) of observed variables (WN: Writing Arabic numbers, sum; RN: Reading Arabic numbers, correct per minute; MC: Magnitude comparison of Arabic numbers, correct per minute; NONDEC: Absolute deviation of slope for linear function of non-decade numbers from 1 (split-half for T1); DEC: Absolute deviation of slope for linear function of decade numbers from 1; ANC: Double-digit additions without carrying: correct per minute; SNC: Double-digit subtractions without carrying: correct per minute; AC: Double-digit additions with carrying: correct per minute; SC: Double-digit subtractions with carrying: correct per minute) for each time point (T1: end of first grade; T2: middle of second grade; T3: end of second grade: T4: middle of third grade)

Latent variable	Observed variable	FL at T1	FL at T2	FL at T3	FL at T4
MuDi	WN	.91	.78	.71	-
	RN	.99	.95	.94	-
	MC	.63	.55	.61	-
Lin	NONDEC	.79 / .88	.90	.84	.92
	DEC	-	.85	84	.85
DD C	ANC	-	.74	.65	.61
	SNC	-	.74	.74	.73
	AC	-	.73	.66	.76
	SC	-	.64	.73	.73

Four out of the five chosen fit indices can be considered as satisfactory for both the unspecific default model and the nested model, indicating that our model fits the data well. In the nested model, individual regression weights between latent variables were trimmed out of the default model based on their significance. In total, 10 non-significant paths were excluded from the default model, resulting in the selected nested model (see Figure 6.3). As a global measure of fit, a comparison between the default and the nested model was carried out by means of a X^2 difference test. The difference in X^2 between the unspecific default model and the nested model is not significant (difference in df = 10; difference in X^2 = 9.6; p = .48). Hence, we will only present results for the selected nested model.

Table 6.4: Fit statistics of the longitudinal model of Study 5

	Selected model (nested model)	Default model (unspecific model)	Saturated model (all paths free)	Independence model (all paths = 0)
Chi-Square (X^2)	459	449	0	3037
Degrees of freedom (df)	349	339	0	406
X^2 / df	1.33 [a]	1.33 [a]	-	7.48
AGFI	.782	.780	-	.141
CFI	.958 [a]	.958 [a]	1.00	0
RMSEA	.048 [a]	.049 [a]	-	.216
AIC	631 [a]	642	870	3095

[a] This value is conventionally considered to indicate satisfactory model fit.

The standardized solution for the selected longitudinal structural model with significant paths only (i.e. regression weights significantly different from zero at $p < 0.05$) can be seen in Figure 6.3.

In general, the high regression weights from the factors at one time point to the same factor half a year later (all but one > .58; all $p < .001$) can be interpreted as a quite stable cognitive developmental progression in our sample.

Importantly, there are some significant regression weights indicating developmental influences between the different latent variables as well. Most importantly, the latent variable MuDi explains a high proportion of variance both of Lin and of DD C. This supports our second and third hypotheses based on the theory of conceptual change, namely that general base-10 understanding as needed for multi-digit number processing supports the acquisition of double-digit calculation procedures as well as of a linear number representation in the number range 0-100. It is important to note that the variance in the factor MuDi is almost exclusively due to performance differences for three- and four-digit numbers, as children presented with substantial ceiling effects in the mastery of double-digit numbers even at T1 as outlined above.

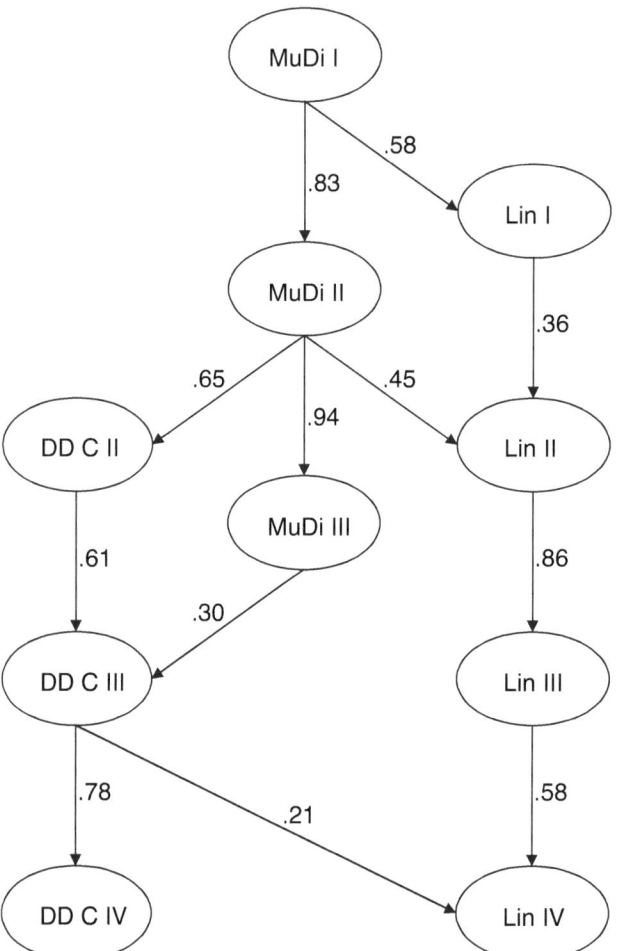

Figure 6.3: Selected nested longitudinal model (standardized solution, structural part only) for linking *multi-digit number processing* (*MuDi*; 3 observed variables each), *double-digit calculation performance* (*DD C*; 4 observed variables each), and *linearity of number representation* (*Lin*; 2 observed variables each) for primary school children from the end of grade 1 (T1) to the middle of grade 3 (T4)

Specifically, MuDi predicts Lin both at the end of first grade (T1) and in the middle of second grade (T2; both $p < .001$), but not at the end of second grade (T3). Interestingly, at T2 the variance of the factor Lin is even better explained by the children's base-10 understanding than by their linearity half a year before, as revealed by a specific test of the null-hypothesis about the equality of two parameters (difference in df = 1; difference in X^2 = 66.6; p < .001).

The impact of MuDi on DD C is significant at T2 (the first time point double-digit calculation performance could be assessed; p < .001) as well as at T3 (p = .005).

Strikingly, we also found a significantly positive effect (p = .016) of DD C at the end of second grade (T3) on Lin half a year later (T4). So, regarding the fourth and last question of this study about the possible causal relations between calculation performance and number representation, our results are in line with the assumption that calculation experience is crucial for a linear number representation, but not vice versa.

6.5 Discussion

The general aim of our study was to investigate possible specific developmental relations between different numerical competencies children are expected to acquire during primary school years. Specifically, we found that flawless transcoding abilities (number symbol knowledge) were not sufficient for a linear number magnitude representation in the same range. Rather, we corroborated our hypothesis that understanding the general base-10 structure of multi-digit numbers fosters the linearity of number representation as well as calculation ability in the range 0-100. Finally we could show that double-digit calculation skills at T3 predicted the linearity of number representation half a year later, and no influence in the other direction was found in our data. In the following, we will discuss these results in more detail.

6.5.1 Is flawless number symbol knowledge sufficient for an exact number representation?

Our first research question was derived from the assumption that the acquisition of a symbolic number code (Dehaene, 2007) directly leads to a linear number magnitude representation in the same range. This hypothesis was mainly based on the neural network model by Verguts and Fias (2004), but never tested empirically yet.

Our results clearly showed that number symbol knowledge per se (flawless transcoding ability) is not sufficient for an exact number representation: Even though our sample presented with considerable ceiling effects in transcoding double-digit numbers at the end of first grade, in about half of the same children (44%) the linear function was not a significant predictor of the number line estimation task (see Table 3). Furthermore, about a quarter of the whole sample was perfectly able to read and write numbers below 100, but did not present with linear estimation patterns in the same range. This means that other cognitive mechanisms than matching symbols to magnitudes must be important for the developmental change to an exact, linearly shaped number representation as well.

6.5.2 Is general base-10 knowledge as needed for multi-digit number processing a predictor for an exact number representation and for calculation skills in the range 0-100?

Employing a longitudinal structural equation model we could show that understanding the base-10 structure of multi-digit numbers larger than 100 (reading, writing, and comparing three- and four-digit Arabic numbers) was highly predictive not only for the linearity of number representation, but also for double-digit calculation performance, thereby lending support to our second and third hypotheses.

The impact of multi-digit number processing on the linearity of number representation in a much smaller range and double-digit calculation performance is not easy to explain, as knowledge about higher multi-digit numbers per se is neither required for double-digit calculation skills nor for an exact representation of numbers in the range 0-100. Yet, the present findings can be accounted for by two developmental models about central conceptual structures in numerical cognition (Case, 1996; Fuson et al., 1997).

Concerning the role of base-10 understanding for double-digit calculation abilities, our results confirm the hypothesis that the comprehension of the decimal system of multi-digit numbers is helpful for understanding arithmetic procedures and therefore the development of multi-digit calculation skills (Fuson et al., 1997). This can be seen as clear support for the important role of conceptual knowledge in understanding the carry- and borrow-procedures needed for multi-digit addition and subtraction (see also Baroody, 2003; Hiebert & Wearne, 1996; Resnick, 1992; Rittle-Johnson & Siegler, 1998).

For the development of an exact linear number magnitude representation, we understand that the comprehension of the relational aspects of natural numbers (as stated by Petitto, 1990; see also Okamoto & Case, 1996) might be the underlying key mechanism why understanding the base-10 system of multi-digit numbers may be helpful. In the developmental model of number representation by von Aster and Shalev (2007), the acquisition of the Arabic number system (and thereby, understanding the numerical relations of the base-10 system) is seen as a prerequisite for the development of a linear number magnitude representation in the form of a mental number line. The multiplicative (or proportional) relation between multi-digit numbers is part of the Arabic number system. As suggested by the ADAPT model (Barrouillet et al., 2004), the transcoding process may be purely syntactic without conceptual understanding of the principles underlying the Arabic number system. Yet, it is plausible to argue that understanding the "meaningful system" of Arabic numbers (Barrouillet et al., 2004, p. 392) is of great help for learning how to transcode or compare multi-digit numbers. According to two developmental conceptual structures models (Case, 1996; Fuson et al., 1997), full comprehension of the Arabic number system corresponds with an integrated bidimensional number concept (see Moeller et al., 2008, for a similar interpretation). This bidimensional number concept is the prerequisite for understanding the relation between two mental number lines – or in other words, the proportional relation between ones and tens, or tens and hundreds; e.g., the distance between 0 and 70 is exactly 10 times larger than the distance between 0 and 7. This view is in line with the recent study by Barth and Paladino (2010) who found that the estimates children give when presented with the number line task are best captured by a proportional function.

To sum up, we believe that the acquisition of the decimal structure of Arabic numbers might trigger the development of an exact, linear number representation because a full comprehension of the Arabic number system comprises an integration of the proportional aspect of the base-10 system into one's number representation – and proportional aspects of numbers can only be captured by an exact, linear representation, but not by a logarithmic (or other non-linear) one.

6.5.3 Is an exact number representation predicting calculation skills or vice versa?

Our last research question concerned the possible developmental influences between the linearity of number representation and calculation performance. In line with the hypothesis formulated by Petitto (1990) as well as recently by Dehaene and colleagues (2008), we found that double-digit calculation performance at the end of second grade (after half a year of training) was a significant predictor of linearity of number representation in the middle of third grade.

We assume that gaining more experience with large calculations for half a year might help children in deepening their understanding of the proportional relations between numbers (which children have already started to acquire by understanding the place-value system of Arabic numbers) and implement them into their numerical representations. Thus, fluency in double-digit calculation might increase the exactness of number magnitude representations even further as already triggered by conceptual understanding of the proportional aspect of the base-10 Arabic number system. This finding that active manipulation of quantities refines the respective numerical magnitude representations fits nicely into the above described conceptual change models (Case, 1996; Fuson et al., 1997), even if not predicted by them in much detail.

6.6 Conclusions

The main result of our study may have important educational implications. In particular, the clear impact of multi-digit number processing on the exactness of number representation in a much smaller number range as well as on double-digit calculation performance has straightforward implications for the planning of future mathematics curricula. Even if domain-unspecific cognitive predictors such as visual spatial abilities (see Study 5) might influence both multi-digit number processing and other numerical capacities, the comprehension of the decimal structure of Arabic numbers is definitely mediating such effects and much easier to train compared to visual spatial abilities or other unspecific cognitive domains. As suggested by Fuson (1990), expanding children's number range to numbers larger than 100 should support their base-10 understanding, because the corresponding number words are regular in most languages. This means that it will be easier for them to uncover the "meaningful structure" of the Arabic number system (Barrouillet et al., 2004, p. 392) as soon as they go beyond the range of double-digit numbers. In our view, this may be especially important in languages in which the number word system is not as transparent as in most Asian languages and thus does not directly support the uncovering of the Arabic number system's place-value structure (cf. Comrie, 2006; see also Study 2).

Therefore, we argue for an earlier emphasis on numbers beyond 100 in mathematics curricula (e.g., in Germany children are currently only expected to understand three-digit numbers at the end of second grade). This might provide children with a better starting point for other important numerical competencies such as exact number representations and multi-digit calculation skills.

7 General Discussion

This thesis investigated the role of multi-digit number processing in the development of numerical cognition. As the acquisition of the Arabic number system is one of the most difficult aspects of numeracy children have to master (Geary, 2000), this research question is of particular importance for educational psychology – which renders it even more astonishing not much is known about the cognitive mechanisms underlying the development of multi-digit number processing and its influence on other aspects of numerical cognition.

Employing mostly correlational methods, we tried to further our understanding of this issue. In five studies using large samples of primary school children from different grades, we gathered evidence for what we think is the core mechanism defining children's maximum performance in a multitude of different tasks - which only have in common that they draw on the processing of multi-digit numbers: namely the conceptual understanding of their base-10 and/or place-value system.

These two aspects are not strictly identical. The term base-10 system indicates that in our number system, we have ten different symbols (number words or digits) which are used to build all larger numbers by means of multiplicative (e.g., "two-hundred") or additive (e.g., "twenty-two") algorithms. In most number word systems, the multiplier have special names or at least morpholocigal markers as well (e.g.,hundred; thousand; -ty for multiplying with ten). On the other hand, the place-value system of Arabic numbers is solely multiplicative in nature, as the position of a digit defines its value (how often it has to be multiplied with ten). The Arabic number system shares the base 10 with our number word system, but theoretically it could be used with other bases as well (like base 60 in the Sumeran and Babylonean number system; see Ifrah, 1981).

In the following sections, we will first discuss how we came to this conclusion based on the findings of the five studies. Next, we will reflect on what our findings imply for the four general research questions outlined in the general introduction, namely whether (i) adult models can be used to describe the development of numeracy, (ii) transcoding in children is a semantic or an asemantic process, (iii) multi-digit number representations are holistic or decomposed in children, and (iv) to which extent visual-spatial abilities explain individual and gender differences in multi-digit number processing.

7.1 The core mechanism underlying children's multi-digit number processing: Base-10 understanding

Several independent findings of the studies presented in this thesis lend support to the notion that the core concept which triggers the development of multi-digit number processing in different kinds of tasks like magnitude comparison or transcoding and in different modalities (e.g., Arabic or verbal) alike is the understanding of "the meaningful structure of the digital forms" (Barrouillet et al., 2004, p. 392), namely its inherent features of the decimal base and the place-value system.

The notion of such a core concept for (at least double-digit) number processing was already brought forwards in the 1990ies by two research groups (Case, 1996; Fuson et al., 1997), but has been largely neglected by cognitive researchers in this domain.

The first and also most important result in favour of a base-10 core concept underlying different kinds of task in different modalities was the finding of a respective cognitive component. Only the assumption of conceptual understanding of the multi-digit number system can explain why children's performance in writing, reading, and comparing Arabic numbers as well as comparing number words should load on one factor (Study 1; see also 7.2 and 7.3). The common variance between all these tasks which was explained by the component multi-digit number processing (as opposed to a calculation component) was independent of highly pronounced gender differences in favour of boys we found in the same tasks (in some cases, effects sizes were even larger than 1; Study 1) – which was our original starting point for the hypothesis that all tasks tapping multi-digit number processing share more aspects than can be explained by contemporary models of numeracy (see 7.2).

Further evidence for conceptual understanding underlying even tasks which are usually thought to be purely asemantic (like writing multi-digit Arabic numbers to dictation; Barrouillet et a., 2004) comes from Study 2 (see 7.3): In a multi-lingual study comparing the transcoding abilities of children speaking a language with inversion of double-digit numbers (from Austria, Germany, and Flanders) to the ability of children speaking a language without this inversion property (from France and Wallonia), we found that inversion showed a general negative effect on transcoding even for items which should be unaffected by the inversion rule (or its incorrect overgeneralization). We argued that this general negative effect of inversion on the transcoding of multi-digit numbers may indicate that inversion should be seen as an irregularity of specific number word systems which acts as an obstacle to understanding the regularity of the whole system. This assumption is in line with the interpretation of the commonly found advantage of children speaking an Asian language with a number word system derived from Chinese (which is much more regular and transparent compared to European number word systems) in a variety of numerical tasks (e.g., Fuson & Kwon, 1991; Miura et al., 1993; Miller et al., 1995; see also Muldoon et al., accepted), namely that these purely multiplicative number word systems foster the acquisition of the Arabic number system because of transparency in the correspondence between both systems compared to the irregular European number word systems.

As all these number word systems (even the ones with inversion) share a base of ten but are different with respect to their employment of additive or multiplicative algorithms, it is likely that the crucial factor in the acquisition of large numbers is less understanding their base-10 but more their place-value system. Study 4 lends support to this assumption, as general visual-spatial abilities were found to predict multi-digit number processing abilities in first graders. Furthermore, we concluded in Study 5 that the mechanism explaining the impact of multi-digit number processing on the exactness of number magnitude representations in the range up to 100 is most likely the growing understanding of the multiplicative (or proportional) aspect of numbers triggered by the acquisition of the Arabic numbers with their inherent place-value system (and later by calculation experience).

To our knowledge, this focus on the key role of understanding the proportional aspect of numbers (which implements their ordinal and cardinal aspects, but implies additional semantic understanding) is relatively new in the field of cognitive research on numerical cognition (but see a recent study by Barth & Paladino, 2010, for proportional estimation patterns in the number line task). Yet, we believe that this thesis could provide evidence for the importance of proportional reasoning in the development of numerical cognition, and that this should be taken into account in future research.

7.2 Adult models are not adequate for the description of the development of numeracy

At least one aspect of both adult models on numerical cognition considered (McCloskey, 1992; Dehaene, 1992; see 1.1 and 2.1.1) is clearly not able to capture performance profiles during the development of numeracy: Both models predict distinguishable components for the processing of verbal and of Arabic numbers, but this could not be observed in a large sample of primary school children. Yet, in all age groups all tasks tapping multi-digit number processing (no matter in which modality) loaded on one single cognitive component. Therefore, the distinction between verbal and Arabic input and output processing systems is not helpful for the description of typical development.

It is important to note that acquisition of the verbal number system starts before the acquisition of the Arabic number system and that mastering the latter will most probably build on the former (von Aster & Shalev, 2007). Yet, our results indicate that reciprocal processes will most likely underlie the development of both, as a common source of variance defines the upper bound of performance in tasks using both modalities throughout the first years of primary school.

Specific predictions that differ between both adult models (McCloskey, 1992; Dehaene, 1992) will be evaluated with respect to our findings in the next two sections.

7.3 The transcoding process in children is at least mediated by semantic number representations

One feature for which the predictions of the two most influential adult models of numerical cognition differ is the nature of the transcoding process: The model by McCloskey (1992) describes the transcoding of numerical input into another modality as mediated by an internal semantic magnitude representation, whereas the Triple-Code model (1992) postulates that transcoding from the verbal to the Arabic modality and vice versa usually happens asemantically. The only developmental transcoding model (ADAPT; Barrouillet et al., 2004) also states that writing multi-digit Arabic numbers to dictation is an asemantic, purley algorithmic process. This model is partly based on an older study by Seron and Fayol (1994) who assessed not only transcoding abilities, but also other numerical tasks in primary school children and concluded that difficulties in writing multi-digit numbers are due to problems in the (purely syntactical) production, but not the comprehension stage of processing.

Yet, as outlined above, we found evidence for semantic mediation of transcoding. If problems in writing Arabic numbers were only due to difficulties in executing the correct algorithms for production, it would be difficult to explain why performance in reading Arabic numbers should highly correlate with performance in writing as found for different age groups in Study 1 as well as in Studies 4 and 5. The algorithms of the two tasks are completely different – their only commonality consists of the underlying meaningful systems. As stated above, also the finding that inversion posed a general obstacle to writing Arabic numbers and not only to items for which this should be expected (Study 2) speaks for the respective importance of understanding the semantic structure of Arabic numbers (see 7.1).

Furthermore, to our knowledge no child has ever been reported with a specific problem in either writing or reading Arabic numbers but good abilities in all other tasks.

We do not think that the execution of algorithms is unnecessary for transcoding, but that the ability to execute these algorithms depends largely on understanding why one should follow them. Actually, a critical evaluation of the study by Seron and Fayol (1994) reveals that their conclusions are largely based on null-results, which could be related to ceiling effects in their data.

Summing up, the results of this thesis are in favour of a semantic transcoding process during development as proposed by McCloskey (1992).

7.4 The decomposed aspect of multi-digit number representation is predominant during childhood

The other crucial difference between the two adult models lies in their conceptualizations of the internal number magnitude representations (see 1.3). McCloskey (1992) proposed a decomposed representation of large numbers relying on their base-10 system, whereas Dehaene (1992) argued for a holistic analogue representation similar to a spatially oriented mental number line. Nuerk and Willmes (2005) reviewed evidence for both types of representation (a hybrid model) at least for the processing of double-digit numbers in adults. Several studies in children have also shown that decomposed processing is at least present in several experimental tasks (Moeller et al., 2009; Nuerk et al., 2004; Pixner et al., 2009), but this does not mean that this type of representation is typically used by children for solving numerical tasks, most importantly magnitude comparison of numbers.

Employing correlational methods allowed us to investigate exactly this question. Our main finding of one cognitive component predicting the performance levels of both two transcoding (which can only be done in a decomposed fashion) as well as two magnitude comparison tasks can not easily be interpreted without taking into account the inherent place-value (or base-10) aspects of multi-digit numbers (Study 1). Furthermore, multi-digit number processing even predicted the linearity (in other words, exactness) of estimates in the number line task, which is thought to directly reflect number representations (Study 5). Hence, we argue that a decomposed internal magnitude representation (McCloskey's view) is at least more dominant than holistic representations (Dehaene's view) in the development of numeracy and maybe even a prerequisite for it.

7.5 The role of visual-spatial abilities in multi-digit number processing

Another yet unsolved question is the importance of visual-spatial abilities for numerical cognition. There are several reasons to believe that visual-spatial abilities are important for the acquisition of multi-digit number processing. One argument is that visual-spatial abilities have been found to mediate gender differences in favour of males in mathematical problem solving, supposedly because of their stronger reliance on visual-spatial as opposed to algorithmic strategies in adults (Casey et al., 1995, 2001; Fennema et al., 1998; Rosselli et al., 2008) as well as in children (Fennema et al., 1998; Gallagher et al., 2000).

It is questionable whether the same would hold for the gender differences we observed in all tasks tapping multi-digit number processing in primary school children (see 2.3.1). Taking into account the respective importance of understanding the place-value system (see 7.1) with its inherent spatial features, the assumption of visual-spatial abilities being important for the mastery of multi-digit numbers is straightforward. Yet, evidence for it is scarce. For example, Geary (1993) suggested a spatial subtype of DD mainly characterized by difficulties in multi-digit number processing based on the finding of spatial acalculia in adult neuropsychological patients (Hartje, 1987; Strang & Rourke, 1985), but in a recent re-evaluation of his originally proposed subtypes he stated that it is not yet clear whether this subtype really exists. On the other hand, a recent study by Zuber and colleagues (2009) found a male advantage in writing Arabic numbers in first graders and that visual-spatial working memory capacity was predictive of transcoding performance. Yet, it is important to note that gender differences in general mathematical problem solving abilities have also been explained by socio-cultural factors such as national gender equality (Giuso et al., 2008; Else-Quest, Hyde, and Linn, 2010) or evaluation of mathematics (Eccles & Jacobs, 1986; Denissen, Zarrett, & Eccles, 2007).

Study 3 was designed to test for the impact of national gender equality on a task tapping multi-digit number processing, namely writing Arabic numbers. Study 4 investigated the relative impact of evaluation of mathematics, visual-spatial working memory capacity, and general visual-spatial abilities on gender (and generally individual) differences in multi-digit number processing.

In Study 3 we could show that in samples of four different countries gender differences in writing Arabic numbers did not correlate with national gender equality, whereas gender differences in subtraction did.

Study 4 revealed that evaluation of mathematics was a significant mediator of gender differences in multi-digit number processing. Visual-spatial abilities also predicted performance in multi-digit number processing, but independent of gender. Visual-spatial working memory capacity was predicted by general visual-spatial ability and had no direct impact on multi-digit number processing.

These results corroborate the psychobiosocial model of gender differences in mathematics (Halpern, 1997), namely that culturally mediated influences such as attitudes and more biologically influenced factors such as visual-spatial abilities are both important for and maybe even interacting in the development of individual differences in numerical cognition.

Therefore, we could show the importance of visual-spatial abilities for the development of multi-digit number processing, but in our study they did not explain respective gender differences. A possible reason for this may be that other visual-spatial tasks which show gender differences themselves even at that young age (e.g., mental rotation) could be more informative in this respect. Another possibility may be that a specific aspect of both numerical as well as spatial cognition, namely proportional reasoning, could be the missing link in this issue. Future studies will hopefully clarify this question.

Another question that remains open is the role of visual-spatial abilities in other domains of numerical cognition, most importantly calculation. In Study 1 we observed gender differences in calculation tasks as well (but less pronounced than for multi-digit number processing tasks and only for one age group; see 2.3.1). Furthermore, in Study 3, national gender equality was related to gender differences in subtraction, but not in multiplication.

Concerning multi-digit calculation, we believe that visual-spatial abilities would have no direct influence, but be mediated by multi-digit number processing, as the longitudinal impact of visual-spatial abilities on multi-digit number processing was only found for the end of first grade, whereas the latter directly predicted double-digit calculation performance even one year later.

In the case of multiplication, it may be speculated that again proportional reasoning could be a candidate for an underlying cognitive process explaining individual differences in performance. Yet, this has to be shown in future studies as well.

7.6 Implications for mathematics education and treatment of developmental dyscalculia

Our main result that multi-digit number processing is not only one important component of the development of numerical cognition, but also predictive for other numerical domains such as double-digit calculation performance and an exact number representation, has important implications for both mathematical education and the treatment of children with developmental dyscalculia (DD).

In this context, it is important to note that all studies presented here tested typically developing children and that concerning multi-digit number processing, performance of most children was ahead of the curriculum (see Study 5).

We therefore conclude that an earlier focus on multi-digit number processing (e.g., at the beginning of second grade) may have positive effects on the development of numeracy in general (see also Fuson, 1990). First, it would provide children with more time for grasping the difficult concepts of place-value and proportionality of number before they will need them for more complex mathematical tasks. Second, it may diminish individual differences in mathematical abilities, as the advantage of children with good multi-digit processing abilities (mostly boys!) for the acquisition of double-digit calculation would be reduced. It may well be possible that low performing children may profit most from such a curricular change, as they are the ones who do not acquire proficiency in conceptual knowledge as fast as good performing children.

Apart from this educational implication, conclusions can also be drawn for the understanding and treatment of DD.

First, we suggest that in future descriptions of DD subtypes, the interplay of verbal, Arabic, and procedural aspects of numerical cognition should be taken into account. A critical evaluation of the subtyping systems described above (see 2.1.2) shows that usually not all necessary information is provided to conclude whether stating a multi-digit number processing-subtype may be more conclusive as compared to an Arabic (e.g., Geary, 2004) or a procedural (e.g., Temple, 1991) subtype. Furthermore, all described verbal DD subtypes are only characterized by problems in counting and/or fact retrieval, but not by transcoding deficiencies, which should be expected if the verbal module as proposed by Dehaene's Triple-Code model (1992) is not well enough developed. In conclusion, we think that a subtype of DD characterized by specific problems in multi-digit number processing in different kinds of tasks and modalities is very likely to exist. Actually, we found a double dissociation between a child showing problems in multi-digit number processing and a child of the same age showing problems in calculation in the standardization sample of the TEDI-MATH (Kaufmann et al., 2009).

If a multi-digit number processing subtype of DD (with problems in multi-digit number processing and calculation) exists, this has straightforward implications for remedial treatment of these children as well. First, children presenting with such a deficit could be earlier identified than by waiting until they show severe problems in multi-digit calculation. Second, specialized cognitive trainings focusing on the regularity of the Arabic number system and the proportional aspect of numbers may be more fruitful than drilling the execution of carry- and borrow-procedures for multi-digit calculation.

Hopefully these predictions will prove to be helpful in the future.

8 References

Ansari, D. (2008). Effects of development and enculturation on number representation in the brain. Nature Reviews Neuroscience, 9, 278-291.

Ansari, D. (2010). Neurocognitive approaches to developmental disorders of numerical and mathematical cognition: The perils of neglecting the role of development. Learning and Individual Differences, 20, 123-129.

Akaike, H. (1987). Factor analysis and AIC. Pschometrika, 52, 317-332.

Ashcraft, M.H., Yamashita, T.,S., & Aram, D.M. (1992). Mathematics performance in left and right brain-lesioned children. Brain and Cognition, 19, 208-252.

Ashkenazi, S., Mark-Zigdon, N., & Henik, A. (2009). Numerical distance effect in DD. Cognitive Development, 24, 387-400.

Bachot, J., Gevers, W., Fias, W., & Roeyers, H. (2005). Number sense in children with visuospatial disabilities: orientation on the mental number line. Psychology Science, 47, 172-183.

Baddeley, A.D. (1986). Working Memory. London: Oxford University Press.

Baddeley, A.D. (2002). Is working memory still working? European Psychologist, 7, 85–97.

Baddeley, A.D., & Hitch, G. (1974). Working memory. In G.H. Bower (Ed.), The psychology of learning and motivation (pp. 47–89). London: Academic Press.

Badian, N.A. (1983). Dyscalculia and nonverbal disorders of learning. In H.R. Myklebust (Ed.), Progress in learning disabilities (Vol. 5, pp. 235-264). New York: Stratton.

Baroody, A.J. (2003). The development of adaptive expertise and flexibility: The integration of conceptual and procedural knowledge. In A. Baroody & A. Dowker (Eds.), The development of arithmetic concepts and skills: Constructing adaptive expertise (pp. 1–34). Mahwah, NJ : Erlbaum.

Barrouillet, P., Camos, V., Perruchet, P., & Seron, X. (2004). ADAPT: A developmental, asemantic and procedural model for transcoding from verbal to Arabic numerals. Psychological Review, 111, 368-394.

Barth, H.C. & Paladino, A.M. (2010). The development of numerical estimation: Evidence against a representational shift. Developmental Science, DOI: 10.1111/j.1467-7687.2010.00976.x.

Becker, K.A. (2003). History of the Stanford-Binet intelligence scales: Content and psychometrics. (Stanford-Binet Intelligence Scales, Fifth Edition Assessment Service Bulletin No. 1). Itasca, IL: Riverside Publishing.

Beery, K.E. (1997). The Beery-Buktenica Developmental Test of Visual-Motor Integration (VMI) with Supplemental Developmental Tests of Visual Perception and Motor Coordination. NJ: Modern Curriculum Press.

Bentler, P.M. (1990). Comparative fit indices in structural models. Psychological Bulletin, 112, 400-404.

Booth, J.L., & Siegler, R.S. (2006). Developmental and individual differences in pure numerical estimation. Developmental Psychology, 4, 189-201.

Booth, J.L., & Siegler, R.S. (2008). Numerical magnitude representations influence arithmetic learning. Child Development, 79, 1016-1031.

Browne, M.W. & Cudeck, R. (1993). Alternative ways of assessing model fit. In K.A. Bollen & J.S. Long (Eds.). Testing structural equation models (pp. 136-162). Newbury Park, California: Sage.

Case, R. (1996). Reconceptualizing the nature of children's conceptual structures and their development in middle childhood. In R. Case, Y. Okamoto, S. Griffin, A. McKeough, C. Bleiker, B. Henderson, & K.M. Stephenson (Eds.), The role of central conceptual structures in the development of children's thought. Monographs of the Society for Research in Child Development (pp. 1-26). Chicago: The University of Chicago Press.

Casey, M.B. (1996). Understanding individual differences in spatial ability within females: A nature/nurture interactionist framework. Developmental Review, 16, 241-260.

Casey, M.B., Nuttall, R., & Benbow, C.P. (1995). The influence of spatial ability on gender differences in mathematics college entrance test-scores across diverse samples. Developmental Psychology, 31, 697-705.

Casey, M.B., Nuttall, R.N., & Pezaris, E. (2001). Spatial-mechanical reasoning skills versus mathematics self-confidence as mediators of gender differences on mathematics subtests using cross-national gender-based items. Journal for Research in Mathematics Education, 32, 28-57.

Cipolotti, L. & Butterworth, B. (1995). Towards a multiroute model of number processing: Impaired number transcoding with preserved calculation skills. Journal of Experimental Psychology: General, 24, 375-390.

Cohn, R. (1961). Dyscalculia. Archives of Neurology, 4, 301-307.

Cohn, R. (1971). Arithmetic and learning disabilities. In H.R. Myklebust (Ed.), Progress in learning disabilities (Vol. 2, pp. 322-389). New York: Grune & Stratton.

Comrie, B. (2005). Endangered numeral systems. In J. Wohlgemuth and T. Dirksmeyer (Eds.), Bedrohte Vielfalt: Aspekte des Sprach(en)tods [Endangered Diversity: Aspects of Language Death]. Berlin: Weißensee Verlag.

Corsi, P.M. (1972). Human memory and the medial temporal region of the brain. Dissertation Abstracts International, 34, 891(B). (University Microfilms No. AA105-77717).

Cronbach, L.J. (1975). Beyond the two disciplines of scientific psychology. American Psychologist, 2, 116-127.

Dehaene, S. (1992). Varieties of numerical abilities. Cognition, 44, 1-42.

Dehaene, S. (1997). The Number Sense: How the Mind Creates Mathematics. New York, NY: Oxford University Press.

Dehaene, S. (2007). Symbols and quantities in parietal cortex: elements of a mathematical theory of number representation and manipulation. In P. Haggard, Y. Rossetti, & M. Kawato (Eds.), Attention & Performance XXII. London: Oxford University Press.

Dehaene, S., Bossini, S., & Giraux, P. (1993). The mental representation of parity and numerical magnitude. Journal of Experimental Psychology: General, 122, 371-396.

Dehaene, S. & Cohen, L. (1995). Towards an anatomical and functional model of number processing. Mathematical Cognition, 1, 83-120.

Dehaene, S., Izard, V., Spelke, E., & Pica, P. (2008). Log or Linear? Distinct intuitions of the number scale in Western and Amazonian Inigene cultures. Science, 320, 1217-1220.

Delgado, A.R. & Prieto, G. (2004). Cognitive mediators and sex-related differences in mathematics. Intelligence, 32, 25-32.

Deloche, G., & Seron, X. (1987). Numerical transcoding: A general production model. In G. Deloche & X. Seron (Eds.), Mathemtaical disabilities: A cognitive neuropsychological perspective (pp. 137-179). Hillsdale, NJ: Erlbaum.

Denissen, J.J.A., Zarrett, N.R., & Eccles, J.S. (2007). I like to do it, I'm able, and I know I am: Longitudinal couplings between domain-specific achievement, self-concept, and interest. Child Development, 78, 430-447.

Desoete, A. (2007). Students with mathematical disabilities in Belgium: from definition, classification and assessment to STICORDI devices. In T.E. Scruggs & M.A.

Mastropieri (Eds.), Advances in Learning and Behavioral Disabilities, (Vol. 20, pp. 181-222). International Perspectives Amsterdam & Oxford: Elsevier Press.

Desoete, A. (2008). Do birth order, family size and gender affect arithmetic achievement in elementary school? Electronic Journal of Research in Educational Psychology, 6 (1), 135-156.

Domahs, F., Delazer, M., & Nuerk, H.-C., (2006). What makes multiplication facts difficult – Problem size or neighbourhood consistency? Experimental Psychology, 53, 275-282.

Domahs, F., Domahs, U., Schlesewsky, M., Ratinckx, E., Verguts, T., Willmes, K. & Nuerk, H.-C. (2007). Neighborhood consistency in mental arithmetic: Behavioral and ERP evidence. Behavioral and Brain Functions, 3:66, doi:10.1186/1744-9081-3-66.

Dowker, A. (2005). Early identification and intervention for students with mathematics difficulties. Journal of Learning Disabilities, 38, 324-332.

Dowker, A., Bala, S., Lloyd, D. (2008). Linguistic Influences on Mathematical Development: How important is the transparency of the counting system? Philosophical Psychology, 21, 523-538.

Ebersbach, M., Luwel, K., Frick, A., Onghena, P., & Verschaffel, L. (2008). The relationship between the shape of the mental number line and familiarity with numbers in 5- to 9-year old children: Evidence for a segmented linar model. Journal of Experimental Child Psychology, 99, 1-17.

Eccles, J.S. & Jacobs, J.E. (1986). Social forces shape math attitudes and performance. Journal of Women in Culture and Society, 11, 367-380.

Else-Quest, N.M., Hyde, J.S., & Linn, M.C. (2010). Cross-national patterns of gender differences in mathematics: A meta-analysis. Psychological Bulletin, 136, 103-127.

Fennema, E., Carpenter, T. P., Jakobs, V. R., Franke, M. L., & Lewi, L. W. (1998). A longitudinal study of gender differences in young children's mathematical thinking. Educational Researcher, 27, 16-31.

Fodor, J. A. (1983). Modularity of Mind: An Essay on Faculty Psychology. Cambridge, Mass.: MIT Press.

Friedman, L. (1995). The space factor in mathematics: gender differences. Review of Educational Research, 5, 22-50.

Fuson, K.C., & Kwon, Y. (1991). Chinese-based regular and European irregular systems of number words: The disadvantages for English-speaking children. In K. Durkin &

B. Shire (Eds.), Language in mathematical education: Research and practice (pp. 211-226). Milton Keynes, PA: Open University Press.

Fuson, K.C., Wearne, D., Hiebert, J.C., Murray, H.G., Human, P.G., Olivier, A.I., Carpenter, T.P., & Fennema, E. (1997). Children's conceptual structures for multidigit numbers and methods of multidigit addition and subtraction. Journal of Research in Mathematics Education, 28, 130-162.

Gallagher, A.M., De Lisi, R., Holst, P.C., Gillicuddy-De Lisi, A.V., Morely, M., & Cahalan, C. (2000). Gender differences in advanced mathematical problem solving. Journal of Experimental Child Psychology, 75, 165-190.

Geary, D.C. (1993). Mathematical disabilities: Cognitive, neuropsyschological, and genetic components. Psychological Bulletin, 114, 345-362.

Geary, D.C. (1996). Sexual selection and sex differences in mathematical abilities. Behavioral and Brain Sciences, 19, 229-284.

Geary, D.C. (2000). From infancy to adulthood: the development of numerical abilities. European Child & Adolescent Psychiatry, 9(2), 11-16.

Geary, D.C. (2004). Mathematics and learning disabilities. Journal of Learning Disabilities, 37, 4-15.

Geary, D.C. (2010). Mathematical disabilities: Reflections on cognitive, neuropsychological, and genetic components. Learning and Individual Differences, 20, 130-133.

Geary, D.C., Bow Thomas, C.C., Liu, F., & Siegler, R.S. (1996). Development of arithmetical competencies in Chinese and American children: Influence of age, language, and schooling. Child Development, 67, 2022-2044.

Geary, D.C., Hamson, C.O., & Hoard, M.K. (2000). Numerical and arithmetical cognition: A longitudinal study of process and concept deficits in children with learning disability. Journal of Experimental Child Psychology, 77, 236–263.

Grégoire, J., Noël, M.-P. & Van Nieuwenhoven, C. (2004). TEDI-MATH (Flemish adaptation: A. Desoete). Brussels: TEMA/Antwerpen: Harcourt.

Grégoire, J., Noël, M.-P., & Van Nieuwenhoven, C. (2005). TEDI-MATH. Test para el diagnóstico de las competencias básicas en matemáticas (Spanish adaptation: M. J. Sueiro Abad & J. Pereña Brand). Madrid: TEA Ediciones.

Guiso, L., Monte, F., Sapienza, P., & Zingales, L. (2008). Culture, gender, and math. Science, 320, 1164-1165.

Halpern, D.F. (1997). Sex differences in intelligence. Implications for education. American Psychologist, 52, 1091-1102.

Hartje, W. (1987). The effects of spatial disorders on arithmetical skills. In G. Deloche & X. Seron (Eds.), Mathematical disabilities: A cognitive neuropsychological perspective (pp. 121-135). Hillsdale, NJ: Erlbaum.

Haugeland, J. (1989). AI: the Very Idea. Boston: MIT Press.

Hecaen, H., Angelergues, R., & Houillier, S. (1961). Les varietes cliniques des acalculies au cours des lesions retro rolandiques: Approche statistique du probleme. Revue Neurologique, 105, 85-103.

Hausmann, R., Tyson, L.D., & Zahidi, S. (2006). The Global Gender Gap Report 2006, World Economic Forum.

Hedges, L.V. & Nowell A. (1995). Sex differences in mental test scores, variability, and numbers of high-scoring individuals. Science, 269, 41-45.

Helmreich, I., Zuber, J., Pixner, S., Kaufmann, L., Nuerk, H.-C., & Moeller, K. (accepted). Language effects on the mental number line: Comparing Italian- and German-speaking children. Journal of Cross-Cultural Psychology.

Hoyle, R.H. (1995). Structural Equation Modelling. Concepts, issues and applications. Thousand Oaks: Sage Publications.

Hyde, J. S., Fennema, E., & Lamon, S.J. (1990). Gender differences in mathematics performance - a metaanalysis. Psychological Bulletin, 107, 139-155.

Hyde, J.S., & Mertz, J.E. (2009). Gender, culture, and mathematics performance. Proceedings of the National Academy of Sciences, 106, 8801-8807.

IEA (1996). Third International Mathematics and Science Study (TIMSS III). TIMSS International Study Center Boston College, Chestnut Hill, MA, USA.

IEA (2007). Third International Mathematics and Science Study (TIMSS 2007). TIMSS International Study Center Boston College, Chestnut Hill, MA, USA.

Ifrah, G. (1981). Histoire Universelle des Chiffres. Paris: Seghers.

Johnson, E.S. (1984). Sex differences in problem solving. Journal of Educational Psychology, 76, 1359-1371.

Jöreskog, K.G. (1969). A general approach to confirmatory maximum likelihood factor analyses. Psychometrika, 34, 183-202.

Jöreskog, K.G., & Sörbom, D. (2005). Lisrel 8.72. Lincolnwood: Scientific Software International.

Karmiloff-Smith, A. (1992). Beyond Modularity. Cambridge: The MIT Press.

Karmiloff-Smith, A. (1998). Development itself is the key to understanding developmental disorders. Trends in Cognitive Sciences, 2, 389-398.

Kaufmann, L. (2002). More evidence for the role of the central executive in retrieving arithmetic facts – a case study of severe DD. Journal of Clinical and Experimental Neuropsychology, 24, 302-310.

Kaufmann, L., Lochy, A., Drexler, A., & Semenza, C. (2004). Deficient arithmetic fact retrieval – storage or access problem? A case study. Neuropsychologia, 42, 482-496.

Kaufmann, L., & Nuerk, H.-C. (2005). Numerical development: Current issues and future perspectives. Psychology Science, 47, 142-170.

Kaufmann, L., Nuerk, H.-C., Graf, M., Krinzinger, H., Delazer, M., & Willmes, K. (2009). TEDI-MATH. Test zur Erfassung numerisch-rechnerischer Fertigkeiten vom Kindergarten bis zur 3. Klasse. Bern, Verlag Hans Huber.

Kimura, D. (2000). Sex and cognition. Cambridge, MA: MIT Press.

Kohr, R.L. & Games, P.A. (1977). Testing complex a priori contrasts on means from independent samples. Journal of Educational Statistics, 2, 207-216.

Kosc, L. (1974). DD. Journal of Learning Disabilities, 7, 164-177.

Krinzinger, H., Grégoire, J., Desoete, A., Kaufmann, L., Nuerk, H.-C., & Willmes, K. (accepted). Differential language effects on numerical skills in second grade. Journal of Cross Cultural Psychology.

Krinzinger, H., Kaumann, L., Dowker, A., Thomas, G., Graf, M., Nuerk, H.-C., & Willmes, K. (2007). Deutschsprachige Version des Fragebogens für Rechenangst (FRA) für 6- bis 9-jährige Kinder. Zeitschrift für Kinder- und Jugendpsychiatrie und Psychotherapie, 35, 341–351.

Landerl, K., Bevan, A., & Butterworth, B. (2004). DD and basic numerical capacities: A study of 8–9-year-old students. Cognition, 93, 99–125.

Laski, E.V., & Siegler, S.R. (2007). Is 27 a big number? Correlational and causal connections among numerical categorization, number line estimation, and numerical magnitude comparison. Child Development, 78, 1723-1743.

Linn, M.C. & Petersen, A.C. (1985) Emergence and characterization of sex differences in spatial ability: A meta- analysis. Child Development, 56, 1479-98.

Liu, O.L., Wilson, M., & Paek, I. (2008). A multidimensional Rasch analysis of gender differences in PISA mathematics. Journal of Applied Measurement, 1, 18-35.

Lonnemann, J., Krinzinger, H., Knops, A., & Willmes, K. (2008). Spatial representations of numbers in children and their connection with calculation abilities. Cortex, 44, 420-428.

McCloseky, M. (1992). Cognitive mechanisms in numerical processing: Evidence from acquired dyscalculia. Cognition, 44, 107-157.

McCloskey, M., Caramazza, A., & Basili, A. (1985). Cognitive mechanisms in number processing and calculation: Evidence from dyscalculia. Brain and Cognition, 4, 171-196.

McCloskey, M., Aliminosa, D., & Sokol, S.M. (1991). Facts, rules, and procedures in normal calculation: Evidence from multiple single-patient studies of impaired arithmetic fact retrieval. Brain and Cognition, 17, 154-203.

Meyer, M.L., Salimpoor, V.N., Wu, S.S., Geary, D.C., & Menon, V. (2010). Differential contribution of specific working memory components to mathematics achievement in 2nd and 3rd graders. Learning and Individual Differences, 20, 101-109.

Miller, K.F., Smith, C.M., Zhu, J., & Zhang, H. (1995). Preschool origins of cross-national differences in mathematical competence: The role of number-naming systems. Psychological Science, 6, 56-60.

Mills, C.J., Abland, K.E., & Stumpf, H. (1993). Gender differences in academically talented young students' mathematical reasoning: Patterns across age and subskills. Journal of Educational Psychology, 85, 340-346.

Miura, I.T., Okamoto, Y., Kim, C.C., Steere, M., & Fayol, M. (1993). First graders' cognitive representation of number and understanding of place value: Cross-national comparisons – France, Japan, Korea, Sweden, and the United States. Journal of Educational Psychology, 85, 24-30.

Moeller, K., Pixner, S., Kaufmann, L., & Nuerk, H.-C. (2009). Children's early mental number line: Logarithmic or rather decomposed linear? Journal of Experimental Child Psychology, 103, 503-515.

Muldoon, K., Simms, V., Towse, J., Burns, V., & Yue, G. (accepted). Comparing Chinese with Scottish children's performance on estimation and other number tasks age 5. Journal of Cross-Cultural Psychology.

Noël, M.P., & Turconi, E. (1999). Assessing number transcoding in children. European Review of Applied Psychology, 49, 295-302.

Nuerk, H.-C., Kaufmann, L., Zoppoth, S., & Willmes, K. (2004). On the development of the mental number line: More, less, or never holistic with increasing age? Developmental Psychology, 40, 1199-1211.

Nuerk, H.-C., Weger, U., & Willmes, K. (2001). Decade breaks in the mental number line? Putting tens and units back into different bins. Cognition, 82, 25-33.

Nuerk, H.-C. & Willmes, K. (2005). On the magnitude representation of two-digit numbers. Psychology Science, 47, 52-72.

OECD (2003). Programme for International Student Assessment (PISA), 2nd Assessment, OECD, Paris, France.

Okamoto, Y. & Case, R. (1996). Exporing the microstructure of children's central conceptual structures in the domain of number. In R. Case, Y. Okamoto, S. Griffin, A. McKeough, C. Bleiker, B. Henderson, & K.M. Stephenson, (Eds.), The role of central conceptual structures in the development of children's thought. Monographs of the Society for Research in Child Development (pp. 27-58). Chicago: The University of Chicago Press.

Opfer, J.E., & Siegler, R.S. (2007). Representational change and childrens' numerical estimation. Cognitive Psychology, 55, 169-195.

Osborne, J.W. (2001). Testing stereotype threat: Does anxiety explain race and sex differences in achievement? Contemporary Educational Psychology, 26, 291-310.

Petitto, A.L. (1990). Development of number line and measurement concepts. Cognition and Instruction, 7, 55-78.

Piaget, J. (1952). The child's conception of number. London: Routledge & Kegan Paul.

Pixner, S., Moeller, K., Zuber, J., & Nuerk, H.-C. (2009). Decomposed but parallel processing of two-digit numbers in 1st graders. The Open Psychology Journal, 2, 40-48.

Power, R.J.D., & Dal Martello, M.F. (1990). The dictation of Italian numerals. Language and Cognitive Processes, 5, 237-254.

Power, R.J.D., & Dal Martello, M.F. (1997). From 834 to eighty thirty four: The reading of Arabic numerals by seven-year-old children. Mathematical Cognition, 63-85.

Raghubar, K.P., Barnes, M.A., & Hecht, S.A. (2010). Working memory and mathematics: A review of developmental, individual difference, and cognitive approaches. Learning and Individual Differences, 20, 110-122.

Resnick, L. (1982). Syntax and semantics in learning to subtract. In T. Carpenter, J. Moser, & T. Romberg (Eds.), Addition and subtraction: A cognitive perspective (pp. 136–155). Hillsdale, NJ: Erlbaum.

Rittle-Johnson, B., & Siegler, R.S. (1998). The relation between conceptual and procedural knowledge in learning mathematics: A review of the literature. In C. Donlan (Ed.), The development of mathematical skills (pp. 75–110). Hove, UK: Psychology Press.

Robinson, N.M., Abbott, R.D., Berninger, V.W., & Busse, J. (1996). The structure of abilities in math-precocious young children: Gender similarities and differences. Journal of Educational Psychology, 88, 341-352.

Robinson, N.M., Abbott, R.D., Berninger, V.W., Busse, J., & Mukhopadhyay, S. (1997). Developmental changes in mathematically precocious young children: Longitudinal and gender effects. Gifted Child Quarterly, 41, 145-158.

Rosselli, M., Ardila, A., Matute, E., & Inozemtseva, O. (2008). Gender differences and cognitive correlates of mathematical skills in school-aged children. Child Neuropsychology,15, 216-231.

Rourke, B.P. (1989). Nonverbal learning disabilities: The syndrome and the model. New York: Guilford Press.

Rourke, B.P. & Finlayson, M.A.J. (1978). Neuropsychological significance of variations in patterns of academic performance: Verbal and visual-spatial abilities. Journal of Abnormal Child Psychology, 6, 121-133.

Rourke, B.P. & Strang, J.D. (1978). Neuropsychological significance of variations in patterns of academic performance: Motor, psycho-motor, and tactile-perceptual abilities. Journal of Pediatric Psychology, 3, 62-66.

Royer, J.M., Tronsky, L.N., Chan, Y., Jackson, S.J., & Marchant, H. (1999). Math-fact retrieval as the cognitive mechanism underlying gender differences in math test performance. Contemporary Educational Psychology, 24, 181-266.

Rubinsten, O. & Henik, A. (2009). DD: heterogeneity might not mean different mechanisms. Trends in Cognitive Sciences, 13, 92-99.

Seron, X. & Fayol, M. (1994). Number transcoding in children: A functional analysis. British Journal of Developmental Psychology, 12, 281-300.

Shevlin, M., & Miles, J.N.V. (1998). Effects of sample size, model specification and factor loadings on the GFI in confirmatory factor analysis. Personality and Individual Differences, 25, 85-90.

Shye, S. (1985). Smallest space analysis. In T. Husen & T.N. Postlethwaite (Eds.), International Encyclopedia of Education (pp. 4602-4608). Oxford: Pergamon.

Shye, S. & Elizur, D. (1994). Introduction to Facet Theory. Thousand Oaks, CA: Sage.

Schumacker, E.R., & Lomax, R.G. (2004). A beginner's guide to structural equation modelling. 2nd. Ed. New Jersey: Lawrence Erlbaum Associates.

Siegler, R.S., & Booth, J.L. (2004). Development of numerical estimation in young children. Child Development, 75, 428-444.

Siegler, R.S. & Opfer, J.E. (2003). The development of numerical estimation: Evidence for multiple representations of numerical quantity. Psychological Science, 14, 237-243.

Sokol, S.M., McCloskey, M., Cohen, N.J., & Aliminosa, D. (1991). Cognitive representations and processes in arithmetic: Inferences from the performance of brain-damaged subjects. Journal of Experimental Psychology: Learning, Memory, and Cognition, 17, 355-376.

Stock, P., Desoete, A., & Roeyers, H. (2009). Predicting Arithmetic Abilities: The Role of Preparatory Arithmetic Markers and Intelligence. Journal of Psychoeducational Assessment, 27, 237-251.

Strang, J.D. & Rourke, B.P. (1985). Arithmetic disability subtypes: The neuropsychological significance of specific arithmetical impairment in childhood. In B.P. Rourke (Ed.), Neuropsychology of learning disabilities: Essentials of subtype analysis (pp. 167-183). New York: Guilford Press.

Temple, C.M. (1989). Digit dyslexia: A category-specific disorder in DD. Cognitive Neuropsychology, 6, 93–116.

Temple, C.M. (1991). Procedural dyscalculia and number fact dyscalculia: Double dissociation in DD. Cognitive Neuropsychology, 8, 155-176.

Thomas, G. & Dowker, A. (2000, September). Mathematics anxiety and related factors in young children. Paper presented at British Psychological Society Developmental Section Conference, Bristol.

Van Garderen, D. (2006). Spatial visualization, visual imagery, and mathematical problem solving of students with varying abilities. Journal of Learning Disabilities, 39, 496-506.

Van Nieuwenhoven, C., Grégoire, J., & Noël, M.-P. (2001). TEDI-MATH. Test Diagnostique des Compétences de Base en Mathématiques. Paris: ECPA.

Verguts, T. & Fias, W. (2005). Neighbourhood effects in mental arithmetic. Psychology Science, 47, 132-140.

Von Aster, M. (2000). Developmental cognitive neuropsychology of number processing and calculation: Varieties of DD. European Child and Adolescent Psychiatry, 9(2), 41-57.

Von Aster, M. & Shalev, R.S. (2007). Number development and DD. Developmental Medicine and Child Neurology, 49, 868-873.

Westermann, R. & Hager, W. (1986). Error probabilities in educational and psychological research. Journal of Educational Statistics, 11, 117-146.

Wilson, A.J. & Dehaene, S. (2007). Number sense and DD. In D. Coch, K.W. Fischer, & G. Dawson (Eds.), Human Behavior, Learning, and the Developing Brain: Atypical Development (pp. 212-378). New York: Guilford Press.

Zuber, J., Pixner, S., Moeller, K., & Nuerk, H.-C. (2009). On the language specificity of basic number processing: Transcoding in a language with inversion and its relation to working memory capacity. Journal of Experimental Child Psychology, 102, 60-77.

9 Appendix

Table A2.1: Pearson-correlations between C-scores of all subtests (MCA: Magnitude comparison of Arabic numbers; MCW: Magnitude comparison of number words; TW: Transcoding: writing numbers; TR: Transcoding: reading numbers; AD: Additive decomposition; S: Subtraction; M: Multiplication; WP: Word problems; AC: Arithmetic concepts) for first grade second semester (1_2; above diagonal) and second grade first semester (2_1; below diagonal)

	MCA	MCW	TW	TR	AD	S	M	WP	AC
MCA	-	.57***	.63***	.63***	.27**	.47***	.47***	.39***	.28**
MCW	.60***	-	.68***	.66***	.21*	.52***	.52***	.40***	.392**
TW	.43***	.54***	-	.87***	.27**	.63***	.51***	.44***	.37***
TR	.60***	.65***	.68***	-	.32**	.58***	.48***	.46***	.34***
AD	.27**	.25*	.33***	.31***	-	.46***	.33**	.42***	.28**
S	.49***	.49***	.46***	.51***	.32**	-	.48***	.52***	.36***
M	.35***	.54***	.52***	.49***	.23*	.56***	-	.60***	.34**
WP	.37***	.42***	.41***	.46***	.31**	.54***	.53***	-	.25*
AC	.32**	.26**	.31**	.34**	.17(*)	.33**	.39**	.45***	-

(*) $p < .10$; * $p < .05$; ** $p < .01$; *** $p < .001$

Table A2.2: Pearson-correlations between C-scores of all subtests (MCA: Magnitude comparison of Arabic numbers; MCW: Magnitude comparison of number words; TW: Transcoding: writing numbers; TR: Transcoding: reading numbers; AD: Additive decomposition; S: Subtraction; M: Multiplication; WP: Word problems; AC: Arithmetic concepts) for second grade second semester (3_2; above diagonal) and third grade first semester (3_1; below diagonal)

	MCA	MCW	TW	TR	AD	S	M	WP	AC
MCA	-	.66***	.65***	.63***	.22*	.33**	.40***	.40***	.35***
MCW	.30**	-	.69***	.62***	.28**	.40***	.40***	.44***	.38***
TW	.43***	.59***	-	.76***	.30**	.51***	.41***	.52***	.37***
TR	.24*	.61***	.64***	-	.35***	.52***	.36***	.43***	.36***
AD	.09	.22*	.25*	.39***	-	.37***	.31**	.33**	.23*
S	.36***	.54***	.58***	.54***	.34***	-	.46***	.42***	.36***
M	.19(*)	.13	.12	.16	.17(*)	.38***	-	.27**	.30**
WP	.12	.31**	.36***	.36***	.41***	.54***	.31**	-	.42***
AC	.26**	.40***	.47***	.42***	.39***	.57***	.37***	.58***	-

(*) $p < .10$; * $p < .05$; ** $p < .01$; *** $p < .001$

Table A3.1: Counts, expected counts, and % correct for each of the five groups (French, Walloon, Flemish, Austrian, and German) and the whole sample, chi-square (χ^2), and levels of significance for each item of Writing Arabic numbers to dictation

Item		French	Walloon	Flemish	Austrian	German	Total	χ^2 / p
4	counts	49	21	46	38	66	220	-
	exp. counts	49	21	46	38	66	220	
	% corr.	100%	100%	100%	100%	100%	100%	
7	counts	49	21	46	38	66	220	-
	exp. counts	49	21	46	38	66	220	
	% corr.	100%	100%	100%	100%	100%	100%	
1	counts	49	21	46	38	66	220	-
	exp. counts	49	21	46	38	66	220	
	% corr.	100%	100%	100%	100%	100%	100%	
11	counts	49	21	46	38	66	220	-
	exp. counts	49	21	46	38	66	220	
	% corr.	100%	100%	100%	100%	100%	100%	
40	counts	49	20	46	38	65	218	4.95 / .292
	exp. counts	49	21	46	38	65	218	
	% corr.	100%	95%	100%	100%	98%	99%	
16	counts	49	21	45	37	66	218	3.32 / .509
	exp. counts	49	21	46	38	65	218	
	% corr.	100%	100%	98%	97%	100%	99%	
30	counts	49	20	46	38	65	218	4.95 / .292
	exp. counts	49	21	46	38	65	218	
	% corr.	100%	95%	100%	100%	98%	99%	
73	counts	47	20	42	35	62	206	1.10 / .895
	exp. counts	46	20	43	36	62	206	
	% corr.	96%	95%	91%	92%	94%	94%	
13	counts	49	20	46	38	66	219	9.52 / .049
	exp. counts	49	21	46	38	66	219	
	% corr.	100%	95%	100%	100%	100%	100%	
68	counts	49	21	46	33	62	211	13.56 / .009
	exp. counts	47	20	44	36	63	211	
	% corr.	100%	100%	100%	87%	94%	96%	
80	counts	45	18	45	37	65	210	8.38 / .078
	exp. counts	47	20	44	36	63	210	
	% corr.	92%	86%	98%	97%	98%	95%	
25	counts	49	21	45	38	65	218	2.08 / .722
	exp. counts	49	21	46	38	65	218	
	% corr.	100%	100%	98%	100%	98%	99%	
200	counts	40	14	28	29	51	162	6.59 / .159
	exp. counts	36	15	34	28	49	162	
	% corr.	82%	67%	61%	76%	77%	74%	
109	counts	45	18	33	29	50	175	7.59 / .108
	exp. counts	39	17	37	30	53	175	
	% corr.	92%	86%	72%	76%	76%	80%	
150	counts	35	17	17	13	37	119	23.62 / <.001
	exp. counts	27	11	25	21	36	119	
	% corr.	71%	81%	37%	34%	56%	54%	
101	counts	45	18	30	31	53	177	11.23 / .024
	exp. counts	39	17	37	31	53	177	
	% corr.	92%	86%	65%	82%	80%	80%	

The role of multi-digit number processing in the development of numerical cognition

700	counts	34	15	24	30	43	146	7.33 /
	exp. counts	33	14	31	25	44	146	.119
	% corr.	69%	71%	52%	79%	65%	66%	
643	counts	31	11	12	13	27	94	15.67 /
	exp. counts	21	9	20	16	28	94	.004
	% corr.	63%	52%	26%	34%	41%	43%	
190	counts	30	14	16	17	33	110	9.49 /
	exp. counts	25	11	23	19	33	110	.050
	% corr.	61%	67%	35%	45%	50%	50%	
951	counts	30	11	11	15	31	98	14.52 /
	exp. counts	22	9	20	17	29	98	.006
	% corr.	61%	52%	24%	39%	47%	45%	

Table A3.2: Counts, expected counts, and % correct for each of the five groups (French, Walloon, Flemish, Austrian, and German) and the whole sample, chi-square (χ^2), and levels of significance for each item of Recognition of unit- and decade-digits

Item		French	Walloon	Flemish	Austrian	German	Total	χ^2 / p
28 [a]	counts	47	12	15	31	51	156	52.89 /
	exp. counts	35	15	33	27	47	156	<.001
	% corr.	96%	57%	33%	82%	77%	71%	
13 [a]	counts	47	12	16	29	49	153	45.37 /
	exp. counts	34	15	32	26	46	153	<.001
	% corr.	96%	57%	35%	76%	74%	70%	
10 [a]	counts	46	9	12	26	44	137	50.99 /
	exp. counts	31	13	29	24	41	137	<.001
	% corr.	94%	43%	26%	68%	67%	62%	
520 [a]	counts	38	6	7	19	33	103	40.27 /
	exp. counts	23	10	22	18	31	103	<.001
	% corr.	78%	29%	15%	50%	50%	47%	
709 [a]	counts	41	8	13	20	42	124	34.11 /
	exp. counts	28	12	26	21	37	124	<.001
	% corr.	84%	38%	28%	53%	64%	56%	
20 [b]	counts	47	13	15	26	55	156	53.45 /
	exp. counts	35	15	33	27	47	156	<.001
	% corr.	96%	62%	33%	68%	83%	71%	
15 [b]	counts	47	14	14	29	56	160	60.29 /
	exp. counts	36	15	33	28	48	160	<.001
	% corr.	96%	67%	30%	76%	85%	73%	
37 [b]	counts	47	14	15	29	56	161	56.71 /
	exp. counts	36	15	34	28	48	161	<.001
	% corr.	96%	67%	33%	76%	85%	73%	
650 [b]	counts	26	2	2	15	19	64	33.19 /
	exp. counts	14	6	13	11	19	64	<.001
	% corr.	53%	10%	4%	39%	29%	29%	
405 [b]	counts	20	1	2	15	9	47	32.18 /
	exp. counts	10	4	10	8	14	47	<.001
	% corr.	41%	5%	4%	39%	14%	21%	

[a] Recognition of unit-digit required; [b] Recognition of decade-digit required.

Table A3.3: Counts, expected counts, and % correct for each of the five groups (French, Walloon, Flemish, Austrian, and German) and the whole sample, chi-square (x^2), and levels of significance for each item of Subtraction

Item		French	Walloon	Flemish	Austrian	German	Total	x^2 / p
4-2	counts	45	20	46	38	61	210	6.88 /
	exp. counts	47	20	44	36	63	210	.143
	% corr.	92%	95%	100%	100%	92%	95%	
9-5	counts	39	21	41	36	65	202	16.40 /
	exp. counts	45	19	42	35	61	202	.003
	% corr.	80%	100%	89%	95%	98%	92%	
5-3	counts	40	20	45	38	61	204	13.90 /
	exp. counts	45	19	43	35	61	204	.008
	% corr.	82%	95%	98%	100%	92%	93%	
6-6	counts	37	21	45	38	63	204	28.67 /
	exp. counts	45	19	43	35	61	204	<.001
	% corr.	76%	100%	98%	100%	95%	93%	
4-0	counts	37	21	45	38	63	204	6.06 /
	exp. counts	45	19	43	35	61	204	.195
	% corr.	76%	100%	98%	100%	95%	93%	
16-4	counts	33	15	39	34	55	176	9.12 /
	exp. counts	39	17	37	30	53	176	.058
	% corr.	67%	71%	85%	89%	83%	80%	
27-6	counts	30	14	34	33	54	165	9.36 /
	exp. counts	36	16	35	29	50	165	.053
	% corr.	63%	67%	74%	87%	82%	75%	
40-20	counts	29	16	35	34	60	174	18.57 /
	exp. counts	38	17	37	30	52	174	.001
	% corr.	60%	76%	76%	89%	91%	79%	
36-10	counts	32	10	19	25	40	126	8.76 /
	exp. counts	28	12	26	22	38	126	.067
	% corr.	67%	48%	41%	66%	61%	58%	
44-26	counts	5	3	3	4	12	27	3.88 /
	exp. counts	6	3	6	5	8	27	.423
	% corr.	10%	14%	7%	11%	18%	12%	

Table A5.1: Stimulus sets for the Reading and the Writing Arabic Numbers tasks at T1 and T2-T4

Reading Arabic numbers		Writing Arabic numbers	
T1 (n=30)	T2-T4 (n=38)	T1 (n=30)	T2-T4 (n=38)
4	81	3	92
7	109	6	103
15	208	14	159
18	728	19	307
81	43	92	160
109	114	103	37
40	260	30	800
208	8000	159	146
728	153	307	513
43	182	160	1600
114	618	37	117
260	1900	800	611
8000	180	146	200
153	4000	70	365
60	503	513	7000
182	312	1600	107
618	104	117	402
1900	536	611	730
180	1300	200	150
4000	112	365	1200
503	400	7000	821
312	780	107	111
104	600	402	920
536	120	730	3000
1300	1037	150	1047
112	6490	1200	6570
400	1003	821	1009
780	1813	111	1613
600	7005	920	3002
120	1061	3000	1308
	4732		3924
	1019		1018
	4509		5204
	1576		1563
	9217		8419
	3052		9063
	1320		1810
	8016		7015

Table A5.2: Pearson-correlations between the observed variables at T1 (end of first grade: above diagonal) and T2 (middle of second grade: below diagonal) for Study 4

	MAQ A	MAQ B	Corsi-Block forward	Corsi-Block back-ward	VMI	VP	Writing numbers	Reading numbers (corr per min)	Magnitude comparison (corr per min)
MAQ A	-	.69***	.12	.16	-.12	.06	.38***	.38***	.30***
MAQ B		-	.05	.06	-.09	.06	.20*	.24**	.14(*)
Corsi-Block forward	.15(*)	.06	-	.45***	.22**	.36***	.09	.06	.20*
Corsi-Block back-ward	.17(*)	.07	.46***	-	.26**	.25**	.24**	.22**	.25**
VMI	-.06	-.11	.30***	.21*	-	.39***	.21*	.15(*)	.22**
VP	-.02	-.05	.32***	.32**	.56***	-	.27**	.17*	.19*
Writing numbers	.33***	.26**	.23**	.30***	.20**	.12	-	.89***	.58***
Reading numbers (corr per min)	.42***	.29**	.12	.25**	.07	.05	.71***	-	.61***
Magnitude comparison (corr per min)	.31***	.26**	.24**	.35***	.20**	.13	.53***	.48***	-

(*) p < .10; * p < .05; ** p < .01; *** p < .001

Table A5.3: Pearson-correlations of observed variables at T3 (end of second grade) for Study 4

	MAQ A	MAQ B	Corsi-Block forward	Corsi-Block back-ward	VMI	VP	Writing numbers	Reading numbers (corr per min)	Magnitude comparison (corr per min)
MAQ A	-	.66***	.07	.16(*)	-.08	.01	.16(*)	.25**	.24**
MAQ B		-	.05	.06	-.09	.06	.16(*)	.11	.20**
Corsi-Block forward			-	.52***	.36***	.20**	.20**	.15(*)	.26**
Corsi-Block back-ward				-	.23**	.24**	.13	.13	.18*
VMI					-	.33***	.09	.07	.13
VP						-	.01	.00	.14(*)
Writing numbers							-	.66***	.51***
Reading numbers (corr per min)								-	.56***
Magnitude comparison (corr per min)									-

(*) p < .10; * p < .05; ** p < .01; *** p < .001

Table A5.4: Pearson-correlations of observed variables (MAQ A: "How good are you at...?"; MAQ B: "How much do you like...?") of all three time points (T1: end of first grade, T2: middle of second grade, T3: end of second grade) loading on latent variable evaluation of mathematics (all $p < .001$)

	MAQ A 1	MAQ A 2	MAQ A 2	MAQ B 1	MAQ B 2	MAQ B 3
MAQ A 1	-	.68	.54	.69	.54	.40
MAQ A 2		-	.67	.54	.79	.48
MAQ A 3			-	.48	.55	.66
MAQ B 1				-	.60	.61
MAQ B 2					-	.58
MAQ B 3						-

Table A5.5: Pearson-correlations of observed variables of all three time points (T1: end of first grade, T2: middle of second grade, T3: end of second grade) loading on latent variable visual-spatial working memory capacity (CB: Corsi Block; all $p < .001$)

	CB forward 1	CB forward 2	CB forward 3	CB backward1	CB backward2	CB backward3
CB forward 1	-	.38	.43	.45	.41	.32
CB forward 2		-	.52	.41	.46	.49
CB forward 3			-	.47	.50	.52
CB backward1				-	.51	.47
CB backward2					-	.39
CB backward3						-

Table A5.6: Pearson-correlations of observed variables (VMI: visual-motor integration; VP: visual perception) of all three time points (T1: end of first grade, T2: middle of second grade, T3: end of second grade) loading on latent variable general visual-spatial ability (all $p < .001$)

	VMI 1	VMI 2	VMI 3	VP 1	VP 2	VP 3
VMI 1	-	.45	.51	.39	.40	.41
VMI 2		-	.58	.33	.56	.34
VMI 3			-	.28	.37	.33
VP 1				-	.38	.43
VP 2					-	.47
VP 3						-

Table A5.7: Pearson-correlations of observed variables (WN: Writing Arabic numbers to dictation, sum; RN: Reading Arabic numbers, correct per minute; MC: Magnitude comparison of Arabic numbers, correct per minute) of all three time points (T1: end of first grade, T2: middle of second grade, T3: end of second grade) loading on latent variable multi-digit number processing (all p < .001)

	WN 1	WN 2	WN 3	RN 1	RN 2	RN 3	MC 1	MC 2	MC 3
WN 1	-	.71	.61	.89	.71	.67	.58	.40	.46
WN 2		-	.85	.62	.71	.70	.39	.53	.51
WN 3			-	.53	.61	.66	.31	.45	.51
RN 1				-	.80	.73	.61	.42	.46
RN 2					-	.83	.48	.48	.50
RN 3						-	.38	.51	.56
MC 1							-	.39	.44
MC 2								-	.57
MC 3									-

Table A6.1: Pearson-correlations of observed variables (NONDEC: absolute deviation of slope for linear function of non-decade numbers from 1 (split-half for T1); DEC: absolute deviation of slope for linear function of decade numbers from 1) of all four time points (T1: end of first grade, T2: middle of second grade, T3: end of second grade, T4: middle of third grade) loading on latent variable linearity of number representation (all $p \leq .001$)

	NONDEC 1a	NONDEC 1b	DEC 2	NONDEC 2	DEC 3	NONDEC 3	DEC 4	NONDEC 4
NONDEC 1a	-	.68	.34	.42	.30	.29	.28	.27
NONDEC 1b		-	.46	.52	.39	.37	.30	.32
DEC 2			-	.76	.59	.56	.45	.54
NONDEC 2				-	.58	.61	.49	.55
DEC 3					-	.71	.61	.49
NONDEC 3						-	.44	.50
DEC 4							-	.78
NONDEC 4								-

Table A6.2: Pearson-correlations of observed variables (ANC: double-digit additions without carrying: correct per minute; AC: double-digit additions with carrying: correct per minute; SNC: double-digit subtractions without carrying: correct per minute; SC: double-digit subtractions with carrying: correct per minute) of three time points (T2: middle of second grade, T3: end of second grade, T4: middle of third grade) loading on latent variable double-digit calculation (all but three p < .001; SC3 x ANC4: p = .001; ANC2 x SC3: p = .002; ANC2 x AC3: p = .008)

	ANC2	AC2	SNC2	SC2	ANC3	AC3	SNC3	SC3	ANC4	AC4	SNC4	SC4
ANC2	-	.64	.53	.42	.36	.22	.43	.26	.33	.28	.35	.32
AC2		-	.53	.55	.47	.46	.41	.33	.42	.32	.42	.35
SNC2			-	.55	.40	.36	.43	.36	.34	.32	.50	.37
SC2				-	.43	.40	.39	.49	.38	.41	.42	.43
ANC3					-	.68	.50	.49	.52	.45	.35	.34
AC3						-	.47	.47	.36	.37	.34	.39
SNC3							-	.46	.41	.51	.45	.34
SC3								-	.27	.36	.36	.48
ANC4									-	.58	.52	.42
AC4										-	.37	.39
SNC4											-	.48
SC4												-

Table A6.3: Pearson-correlations between observed variables at T1 (NONDEC: absolute deviation of slope for linear function of non-decade numbers from 1; all p < .001)

	Writing numbers (sum)	Reading numbers (corr per min)	Magnitude comparison (corr per min)	NONDEC (split-half odd)	NONDEC (split-half even)
Writing numbers (sum)	-	.89	.58	-.48	-.47
Reading numbers (corr per min)		-	.61	-.48	-.43
Magnitude comparison (corr per min)			-	-.41	-.35
NONDEC (split-half odd)				-	.68
NONDEC (split-half even)					-

Table A6.4: Pearson-correlations of observed variables (WN: writing Arabic numbers, sum; RN: reading Arabic numbers, correct per minute; MC: magnitude comparison of Arabic numbers, correct per minute; NONDEC: absolute deviation of slope for linear function of non-decade numbers from 1; DEC: absolute deviation of slope for linear function of decade numbers from 1; ANC: double-digit additions without carrying: correct per minute; AC: double-digit additions with carrying: correct per minute; SNC: double-digit subtractions without carrying: correct per minute; SC: double-digit subtractions with carrying: correct per minute) at T2 (middle of second grade; above diagonal) and T3 (below diagonal)

	WN	RN	MC	NONDEC	DEC	ANC	AC	SNC	SC
WN	-	.71***	.53***	-.58***	-.53***	.31***	.41***	.41***	.39***
RN	.66***	-	.48***	-.45***	-.42***	.35***	.45***	.39***	.47***
MC	.51***	.56***	-	-.39***	-.35***	.34***	.45***	.35***	.35**
NONDEC	-.39***	-.28**	-.33***	-	.76***	-.26**	-.28**	-.35***	-.39***
DEC	-.37***	-.33***	-.29***	.71***	-	-.15$^{(*)}$	-.28**	-.30***	-.35***
ANC	.33***	.36***	.39***	-.17*	-.16$^{(*)}$	-	.64***	.53***	.42***
AC	.40***	.45***	.32***	-.18*	-.13	.68***	-	.53***	.55***
SNC	.34***	.42***	.47***	-.22*	-.20*	.50***	.47***	-	.55***
SC	.35***	.45***	.33***.	-.24**	-.24**	.49***	.47***	.46***	-

$^{(*)}$ p < .10; * p < .05; ** p < .01; *** p < .001

Table A6.5: Pearson-correlations of observed variables (NONDEC: absolute deviation of slope for linear function of non-decade numbers from 1; DEC: absolute deviation of slope for linear function of decade numbers from 1; ANC: double-digit additions without carrying: correct per minute; AC: double-digit additions with carrying: correct per minute; SNC: double-digit subtractions without carrying: correct per minute; SC: double-digit subtractions with carrying: correct per minute) at T4 (middle of third grade)

	NONDEC	DEC	ANC	AC	SNC	SC
NONDEC	-	.78***	-.24**	-.27**	-.05	-.17*
DEC		-	.22**	.28**	.08	.24**
ANC			-	.58***	.52***	.42***
AC				-	.37***	.39***
SNC					-	.48***
SC						-

Acknowledgements

My dearest thanks go to...

- Klaus Willmes, who was a "Doktorvater" in the best sense. He always had time for support when needed and could answer almost every question, no matter what it was about. At the same time, he facilitated my independent working and let my ideas evolve freely. I really hope we will work together in the future.

- Liane Kaufmann, who aroused my interest in developmental neuropsychology and opened the door to science for me.

- Guilherme Wood, who taught me how to use Structural Equation Modelling and was of great help in specifying all presented models.

- Kerstin Konrad, who made me change my mind about my future in Aachen.

- H.-C. Nuerk for insights in how experimental psychology is done properly.

- Martina Graf for organizing data collection for the TEDI-MATH in Germany.

- Jacques Gregoire and Annemie Desoete for providing the French, Walloon, and Flemish TEDI-MATH data.

- Katharina Wopp, Friederike Feldmann, and Jan Lonnemann for help with the data collection for the longitudinal study.

- All former and current colleagues for interesting discussions and lunch breaks.

- My family and friends for making my non-scientific life interesting and valuable.

- All children and teachers involved in the presented studies: Whithout them, developmental research would be impossible.

I want morebooks!

Buy your books fast and straightforward online - at one of world's fastest growing online book stores! Environmentally sound due to Print-on-Demand technologies.

Buy your books online at
www.morebooks.shop

Kaufen Sie Ihre Bücher schnell und unkompliziert online – auf einer der am schnellsten wachsenden Buchhandelsplattformen weltweit! Dank Print-On-Demand umwelt- und ressourcenschonend produziert.

Bücher schneller online kaufen
www.morebooks.shop

KS OmniScriptum Publishing
Brivibas gatve 197
LV-1039 Riga, Latvia
Telefax: +371 686 204 55

info@omniscriptum.com
www.omniscriptum.com

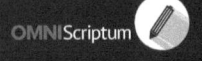

Printed by Books on Demand GmbH, Norderstedt / Germany